SOLAR RETURNS

The simplest method of timing

Author: Bern Jurečič
Title: SOLAR RETURNS
The Simplest Method of Timing

Year of publication: 2024
Imprint: Independently published
Publisher: Kindle Direct Publishing (KDP), Amazon
Copyright: © 2024 Bern Jurečič. All rights reserved.
Design: Janja Baznik
Charts: SolarFire (Astrolabe Inc.)
Cover Image: Adobe Stock

Slovenia, August 2024

Bern Jurečič

SOLAR RETURNS
The simplest method of timing

Table of Contents

Introduction 7
On Solar Returns 12
Story About a Year 43
Elements of the Solar Return Horoscope 67
Timing 133
Stargazer's Cookbook 157
Lunar and Planetary Returns 192
About our Perception 214
Solar Returns in Practice: Margaret Thatcher's Battles
and Victories 226
Case Study: Whatever Happened to Paul 255
Where to Celebrate a Birthday 277
Conclusion 292
Bibliography: 295
Appendix 1 297
Appendix 2 299

I
INTRODUCTION

Another book on the solar return interpretation, which I bought quite a few years ago in order to master this astrological prediction technique, starts with a very self-confident statement: *'In a way, this book is a gift to myself. It was written at the tenth anniversary of my active involvement with astrology."* I note here my modesty, because my book on this field of astrology comes after almost thirty years of intensive work, mainly focusing on this technique. Moreover, as evident further in the book, I myself have spent a whole decade and more just to gain sufficient confidence to produce this book - even after I had already used, tested and proved the technique in dozens, hundreds of practical examples in my work with clients.

The book, I mentioned at the beginning, is based on extremely thorough research; it is elaborate and full of various important calculation formulae that are essential for the template interpretation of solar returns. In my book, however, you will not find so much theory. What it does offer is a view of the yearly horoscopes that is not common today, and what is more, it even describes the path to the surprising finding that the technique, I have been using with great satisfaction for many years in my astrological work on an almost daily basis, is not an ordinary technique, but simply **something more**. In fact, over the years of working in this field, I have come to the realization that this technique has been badly neglected, but also to the belief that there is no reason for this, as well as to the intention to clarify and correct this matter. All this simply because the solar return technique - a much more useful and simpler method than we believe - deserves it.

*

Every human invention, product and result has a story to tell, a narrative of how the idea came about, along with the relating circumstances and obstacles. The present book is no different. Although we are witness to mass production of publications nowadays, some books take a longer period of time and have quite a history. The present book is one of these. Its story of origin is not quite typical, and neither is my astrological journey.

At the beginning of my "more serious" involvement with astrology, I met a remarkable girl, a student called Diana, at an educational seminar, organized by the Croatian journal *Astro magazin*, whom all the participants, including the lecturers, quickly recognized to be far ahead of the rest, one might even say "extraordinary", in her understanding and handling of astrology.

Later, during the war, which took her away from home and her studies, we worked together intensively for several years, which for me was an invaluable learning experience. The work mostly took place at weekends and was very tiring, especially because at that time there were no computer programs that could draw an astrological chart in a circle - I used to draw charts for clients from the astrological chart matrix before weekends, and Diana drew many of them during astrological sessions with clients ...

The most striking feature of her work was the–ease with which she used progressions and especially directions (solar arc directions). But most of all she used examinations of current and the next couple of years in the lives of our clients, i.e. the »yearly horoscopes«, which were left to me for use and further development some time later, when our excellent collaboration had come to an end.

SEARCH

But there is a second stage in the making of this book, which is even more interesting. After a few years of using this technique, which gradually has become my main predictive technique, I reached a point, where I needed to find a better tool for calculating natal charts. Up to that point I had been using a simple American program called *ZodiacSigns*, which was still a DOS program. This little program did not, of course, allow for the wondrous variety of calculations of today's software, but only a basic calculation and the insertion of an additional horoscope into chart, which made it possible to process the synastry and all the aforementioned horoscopes, except that quite a lot of the calculations still had to be done by hand.

I came across an interesting website by a Spanish astrologer called Javier, offering an interesting enough programme that I decided to contact the author. When I received the demo program and tested it, I was surprised to find that the option to calculate solar returns was not among the buttons opening the individual modules. I emailed the author to ask whether it was possible for me to have missed such a function, or whether there was in fact no module for calculating solar returns. I received a short and clear answer that Javier and his co-author had not made such a module at all, claiming that "solar returns didn't work anyway". I was rather disappointed and even a bit offended *("How the hell do they not work when all my consultative astrological work is based on solar returns?!")*, but I obviously did not buy the program, having found it useless.

Sometime later I met another world-famous colleague from abroad, who was only barely familiar with solar returns and had not used the technique at all, and yet also claimed that "according to him, this technique didn't work at all«. When I explained my principle to him, he flatly replied that he had never heard of it.

As a result, I studied solar returns with dedication and found that most sources only discussed them in general terms; however, I did not find any information about the timing within the year anywhere. Moreover, I was now looking from a completely different perspective at the fact that most of the astrology textbooks, that had piled up in my library over the years, were only mentioning solar returns in passing reference or even omitted them in some cases. So after having checked dozens of websites, where solar returns were at least mentioned, I finally realized that my daily astrological routine was actually "something else", something new. But since a personal horoscope "works" and humans being just humans, Saturn squatting on my Ascendant did not allow me to be euphoric - the disbelief that this approach could be pioneering was too strong. Time and again, I was only receiving astonished looks.

One of the few exceptions was one of the most eminent European astrologers, the late Bruno Huber, whom I met at the Astrological Congress in Lucerne in 1996 and interviewed for a Slovenian magazine. In a brief astrological conversation after the interview, we also touched the subject of solar returns, and when I asked him if this was the right technique (how typical of a Virgo!), he succinctly answered: *"I don't know of such a technique, but if it works, it exists. And that's what we need to keep working on!"*

INTRODUCTION

"A COMPLEMENTARY TECHNIQUE ONLY"

After many years of hard work on thousands of cases to prove the accuracy of this method I got the final confirmation, which the god of time Saturn took long to bring about; it took place at the International Congress in Basel in 2006, where I bought the latest edition of the world-famous book on the solar return technique by the American astrologer Ray Merriman. I quickly found in this source a similar explanation of the solar return as I had found so many times before, namely that "a planet works sometime in the year". Then, two days later, at a panel discussion on forecasting techniques in astrology, I posed a question about solar returns to two prominent astrologers, who both replied that they were using this technique (the other three did not mention it, having used other techniques). I received the same answer from both of them, namely that "a solar return horoscope can at most provide us with additional and complementary information to other charts and forecasting techniques", a view I was already familiar with from the literature. At this point it was finally clear that I myself had been practising something quite different for years.

I didn't start working on the present book straight away, because it took a long time to gather and check the material; besides, many parallel activities have been going on, so it really took a long time to complete the work.

The book is not only supposed to prove the efficiency of **the technique** – as evidenced not least by the thousands of conversations and archived data, accumulated over several decades - but also to clarify **why** it works the way it does, i.e. on what basis.

II

ON SOLAR RETURNS

It goes without saying that the modern time is already completely saturated with computers and the various benefits resulting from their applicability. These miraculous cabinets and boxes do gigantic jobs for us that, not so long ago, took hours of work and litres of sweat; they help us organize our lives and are often accused of being smarter than us and of hindering us, yet in reality we cannot even imagine surviving without them nowadays. If the whole system were to collapse for some reason, many people would no longer be able to open their fridge, a good part of global traffic would stop, traffic lights would go out and we would be left without most of the information available to us today. With the recent expansion of social networks a major computer blackout would also kill off virtually all our friends in an instant.

That is why libraries have less and less visitors these days, and even newspapers are slowly falling out of favour, because everything is going on in the world - in pictures, words and commentaries – can be found online, and very often on *Facebook*. Thus, at the beginning of 2012, the news of the sudden death of the American singer Whitney

Houston was announced. Hers is a well-known story, her life having filled the columns of the yellow press and chronicles around the world for a number of years. It is no wonder, therefore, that when I first heard of her death in the *Astrology* Facebook group, theories of suicide or a drug overdose immediately emerged. Especially, of course, due to the fact that she was found drowned in the bathtub of the hotel room, where she and her team were preparing for the Oscar award ceremony. The notes from the group members prompted me to look at her horoscope, because the suicide story did not seem plausible to me at all. I looked up her place of residence on the Internet and did the calculations. My predictions slowly took shape, and on 12 February at 12.07 I joined the conversation and wrote the following: *"Having studied her solar return, I am curious to know what will eventually be announced as the cause of her death. This year has not been depressing or bad for her. She had a very uncomfortable period between March and July 2011, and during this year I think she was on the upswing, in a kind of harmonization period (with herself). There were changes for the better and plans on the horizon. Not very concrete ones, however, but certainly wishes to get back on a stable track – I won't be too surprised if it turns out that she died of cardiac arrest in a bathtub."* In addition, I posted the following on my profile: *"The picture shows her last solar return, of which she only lived through the first half, and the first halves of some of the houses look anything but depressing. It is quite likely that she was about to make a comeback, she had a new start, her popularity was again on the rise ... quite a lot of this, in fact. Unfortunately, it turned out differently. - - - - May she go in peace." (posted on fb 13 February 2012)*

But I was even more surprised the next day, because in the same *Facebook* group, a member circulated an article, published in the Croatian press on the morning of that day, mentioning the singer's last interview, which was recorded for *Access Hollywood* magazine in December 2011, where she said, among other things, that she was "looking forward to the years to come", that she had a lot of support, not only from the close family, especially her mother and her daughter, but also from the "thousands of fans who were cheering her on". She left the journalist with the impression that she was preparing to re-enter the big stage, working on a new film, planning to record a new album.

The cause of her death was soon announced a combination of sedatives and alcohol, resulting in drowning in a bathtub, which also rules out suicide, the starting point of the thesis.

As a footnote, my notes, published before I learnt about this interview, corresponded much better to the actual situation than the assumptions written in forums and based on the prejudice, that the late singer "helped" generate with her turbulent life.

*

Two weeks before, in January 2012, it was also reported that the New Zealand police, in cooperation with the FBI, had arrested Kim Schmitz, a German national, who had been a resident of Auckland for the last four years, on hacking charges (although the official reason for the arrest was suspicion of money laundering). The question of how this affair would turn out was raised in the above-mentioned forum. My view on Mr Schmitz's solar returns is published in a few short notes, as follows in free translation: *"One small problem is that Kim Schmitz was caught practically on his birthday, which may make it difficult to analyze the solar return chart. ... Pluto is at 08°02 and MC is at 08°07, which is almost the same, and yet a little before the solar return itself! In the ninth house (Pluto vs. freedom, but also a foreign country), and at the same time within reach of the tenth house (status, i.e. meaning »a serious threat to status just before the birthday«). "In the solar return there is Saturn in the seventh house, placed within two months after the birthday, which means that things won't work out then (too strong enemies), especially the attempt just before (natal Uranus just before this Saturn). But Jupiter is in the first house some two and a half months after the birthday, and this placement will already bring freedom (first house and Jupiter = me free?). Let's hope."* (published 20 Jan. 2012)

Despite media (and undoubtedly political) pressure and the consequent belief that he would not escape a long prison sentence, Kim Schmitz was out of prison after only a month; a few weeks later he was already in control of his defence and planning his activities in freedom.

*

In an astrological consultation an acquaintance was talking about moving to California, where her son was to finish high school. It was a major move, as the client was not American but Slovenian, so it was a move to another continent. Some basic information was given, but the question was whether or not the move would be successful. The astrological charts made me doubt her timetable, which was to have this project completed in the second half of September of that current year; I said that I thought it more likely that the move would take place as early as the end of July. *'That will not be possible,'* she said, because she would not even be in the USA with her son at that time. I threw up my hands and insisted on my assumption; I further assumed that this move could be carried out by luck and rather quickly, whereas she herself was expecting a few complications.

In our next conversation at the end of that year (via *skype*, as she was already living in the USA at the time), she told me that in the second half of July, an ideal opportunity suddenly turned up, and she and her son grabbed it, flew across the ocean, made the arrangements and completed the action from meeting to viewing and signing the contract in one day! And the move itself was actually completed on the first of August!

*

One of my clients, let's call her Angela, came for a consultation in the middle of one summer on the recommendation of a friend. When we parted, I wrote down a note that I did not know what this lady had come for, because we had somehow talked past each other, not having been able to find the right wavelength for a whole hour and a half. Even her gratitude for my effort and information had not been enough to make me feel good about the conversation. We discussed her demanding job and unpleasant working environment, a possible change of job, and the partnership that had ended two months before the consultation.

When she called again less than a year later and asked for another astrological consultation, I took it as a usual routine, but my surprise came as soon as I took her case out of the archives and read my note. I had been anticipating her visit with excitement, wondering about her motives for another consultation.

She told me as soon as she arrived that she had been urged to do this by *"some forecasts that unfolded exactly as predicted"*, in particular that *"in four months' time she would meet a new person and would already be thinking about living together"*, because, as she said, *'I'm not like that at all, things are much slower with me"*. She thought it was completely impossible. But that's exactly what happened: she met a man in December, and after a month they were already living together, despite the fact that her birth horoscope is really very fixed (Sun in Taurus and most of the planets in fixed signs).

*

Four pictures, four cases. Astrologers have many approaches and techniques at our disposal to give us insight into events and the realization of birth potentials, but as a rule two of the most frequently used techniques are secondary progressions and transits. These give astrological predictions (or "forecasts", as the term is more commonly used) enviable accuracy, if only the right and reliable data are available. In all of the above examples, three of which are fresh and not specifically selected from the endless array of notes, consultations, predictions and interpretations that a practising astrologer accumulates over decades of consultation with clients, similar "results" of astrological analysis could probably have been arrived at by a variety of approaches - the range of astrological techniques is of course much greater - but I arrived at them exclusively by the use of "yearly horoscopes", more correctly, solar return horoscopes. Not only can information about events be gained from solar returns, but timing is also possible, as demonstrated above by the very concrete examples. This puts solar returns, an astrological technique somewhere on the fringes of the list of popular predictive methods, into a new and different perspective. Their applicability is greater and more practical than most records and current astrological theory on predictions would suggest.

The present book provides insight into the use of solar return horoscopes in our everyday work with clients.

A MOMENT IN TIME

The word "horoscope" is derived from the Greek word *horoskopos*, which is composed of the words *hora* (hour, season, period of time) and *skopos* (observer, observed), also *skopein* (see, observe, reflect). In simple terms, a horoscope is therefore a picture and a description of time. A little more precisely, it describes the qualities and potentials of a moment, a fragment of time, which for some reason is important enough for us to draw our attention. This moment can represent anything, be it a birth of a person, a country, a company, a beginning, end or some important intermediate stage of a relationship, of a development, or a moment when a dilemma, a question arises. There is no limit to meaning of a moment. So every car on the road, every chair, every orange, thought, traffic accident, piece of information has its 'moment', we might even wonder when we first saw a black cloud in the southern sky – but who cares about the horoscope of the leftmost chair in the third row?!

And yet it too has its own horoscope, just like its neighbour or every car in the line of cars coming out of the factory just a few minutes apart. Because of the small time lag they have quite similar horoscopes, and yet each one has its own. Just as in the case of a human being, when twins are born a few minutes apart.

But why be interested in the beginning? Couldn't we be interested in any moment in time, a moment in life of a person, a country, a chair? Well, we are interested in these too, but these in-between moments describe a temporary and transitory state of an entity; they do not identify it because it is already conditioned by the basic information, which is – its beginning. For a human being such a primary information (certainly one of the most crucial and used) is his birthday. It was already in use in earlier times [1], but since this information was not always available or accurate, it was supplemented with various other data, such as the name of the father, the colour of hair and eyes, address, and so on, in order to identify and locate the person as much as possible. In recent decades, a unique civil registration number has been introduced, which contains the date of birth in its first seven digits, and this information is used in most records and routine certificates, applications or various other documents.

Generally speaking, the 'beginning of something' is always highlighted. The older the brewery, the more shining the founding year on the product label. Most countries in the world celebrate their 'birthday' (or some other day intrinsically linked to their founding, independence and so on). And last but not least: when buying a second-hand car, in most cases the first thing to ask for is the year of its production. Surely one could also ask about the colour, engine capacity, equipment, last service, fuel consumption... but that usually comes later. In some ways, age tells us more about the car than any

---

1   Let us not forget that in the Christian world, until a century and a half ago, instead of a date of birth, the date of baptism was written in registers. This was done by arrangement, often depending on the distance from the church or priest, the weather, a family illness or some other obstacle, but it was written down. But even this is a kind of beginning – from that day on the child belonged to the communion of the Church ... For identification in a society where Christianity prevails, this date is just as valid as the date of birth. However, there is necessarily a dilemma as to whether or not an astrologer can use such a date. If no other time is available, it is certainly better than nothing, since it is written down, church records generally being reliable. Several such examples can be found in places where, for some reason, there was an interest in recording a date other than the real one. For example, in our part of the world (and especially in the countries of the former Yugoslavia), births in the last days of the year were often recorded as "1st January" of the following year - so that the child would be subject to compulsory military service one year later.
In these cases, of course, it is the date that is concerned, but the same is true of the time of birth. It is not uncommon for a person to have two dates, one written down (possibly from a few decades ago, when birth times were - if ever - recorded to the nearest hour or half-hour), and the other, saved in the mother's memory. And cases when the mother's information turns out to be more accurate, are not even rare...

other piece of information, because it is often tied to part of our knowledge, based on experience – even if only a prejudice - and we formulate the following questions depending on the age question. If we don't like the information, we might even stop asking, being no longer interested in the purchase...

Although it is possible to calculate a horoscope for anyone or anything if one knows the starting data, i.e. the place and time of the beginning, the rest of the story on solar returns will however be more focused on people, i.e. the solar returns ('yearly horoscopes') of people. This is the direction in which the rest of this text is set, and the reader should keep it in mind that both, the natal horoscope and the solar return, can also apply to other identities.

THE MOMENT AND THE HOROSCOPE

The first personal horoscopes were calculated in the time of ancient Greece, although various examples can be found in literature, where authors claim to have been "the first". The basic premise of Hellenistic astrology is that the birth horoscope of an individual constitutes the basis, the starting point for any predictive technique. A birth chart is therefore primarily a picture of potential, not a picture of character or personality. It shows the qualities that a person has brought into the world, which are conditioned by many influences, from genetics, karmic history, the circumstances of birth, the pregnancy of the mother and more. A birth horoscope therefore is a kind of a seed that later evolves into a description of personality development. A more modern metaphor might be that a birth horoscope is a kind of DNA of a person - a starting point (a starting position, a blueprint, an initial potential) for everything the person is to experience in life. Understanding a birth chart therefore opens up complex information about the future of the respective person. Predictive techniques are therefore only a kind of a footnote (and at the same time an upgrade) to the birth chart, as they only determine **when** in life the written potentials will unfold.

"I will never forget this moment."

We have all heard that phrase in our lifetime, and it speaks about the impression of something very important. But it is the important events that have the greatest influence on our life path, decisions and development, and these are the ones that we will find marked in our horoscope. This is where we encounter the concept of relativity, because the question is what is written down, or, to go one step further: what is important and what is not. We are more likely to remember events that have hurt, threatened and upset us, while the fact that we have escaped from an unpleasant situation rather fortunately is soon and easily lost from the memory of many people.

To illustrate what is and what is not important, think of a situation, where carelessness causes milk to spill on the stove. No big deal, we will say, clean up the mess, and maybe even open a window and air out the kitchen. Could that be worth a notation in the horoscope? Almost certainly not, it is too unimportant. But if that boiling milk happens to spill over our young child, who suffers some consequences, even if only short-term, the incident will suddenly turn into a much more significant event, and it is quite possible that we will be able to detect it in the horoscope.

If we know, when the event we are interested in happened, we have the possibility to study its quality by calculating a horoscope for the moment it took place. Each of the elements of a horoscope – however many we include – relates a piece of the story, and together they form a mosaic of information that can be rounded off into a story, or used to predict what will happen next, for example, how a story will unfold or what will happen to the person whose horoscope was cast for that moment; the possibilities are endless.

The most literal illustration of this principle is a horary chart, i.e. an astrological chart calculated for the moment, when a question is posed that the querent is interested in enough to expect an answer. This chart is not bound to a person nor to one's potentials, it is only limited to everything that concerns the question. More closely tied to natal chart are astrological charts of transits, solar arc directions, progressions and all the other techniques we use for astrological prediction; if we are to interpret what they are supposed to bring to the person in question, we need to know that person's potentials and then compare the charts.

This is also the case with a solar return horoscope. It shows the quality of the moment, for which it is calculated, and, in line with the experience that the information we seek is hidden in the moment, a solar return horoscope provides information about the whole year, starting at that exact moment. Such a horoscope has its own "expiry date", this being the horoscope we will calculate for the next birthday, more precisely for the very moment of the next birthday, when the Sun will return to exactly the same position it occupied when we were born.

WHY SOLAR AND WHY RETURN

As the name of this astrological technique suggests, it is the "return of the Sun". Or as Sheila Geddes puts it in her textbook *The Art of Astrology*: "*The solar return is the time of year, when the Sun reaches the exact position it occupies in the birth chart.*" As reader is probably already aware from the literature or from knowledge acquired before reading this book, there are several types of returns, including lunar, Jupiter and other returns, all of which are derived from the study of the cycles of the planets and other celestial bodies, which are the foundation of astrological theory and practice. If we take as a basis a simplified sketch of the solar system, in which the planets – from Mercury onwards - revolve around the Sun in ever larger circles, or orbits, we can imagine that after a certain period of time (called the orbital period) each of the planets returns to its starting position, meaning it finds itself in the same position it occupied at the beginning of our observation.

If we imagine the solar system as a circle drawn on a piece of paper, i.e. two-dimensionally, it is quite possible that an element is placed at the exact leftmost position at any given moment. After the orbital time, calculated for the motion of such an element or planet, we can calculate the exact time of its reappearance. And in the case of Jupiter, this will occur after 11.86 years, because that is its orbital period; in the case of the Moon, it will take place again after about 29 days, and in the case of the Sun, after one year. A 'Year' is a unit, derived precisely from the orbital period of the Sun. A Solar return is therefore an astrological technique, based on the Sun returning to the position that is (astrologically) exactly the same as the position it occupied at the exact moment of the birth of the person in question. The Sun returns to this exact degree once a year, approximately on the birthday of the person. [2]

This means that a solar return horoscope is arrived at in the reverse way to a birth horoscope. In the latter, the starting point is the date, i.e. the calendar representation of the moment of birth, on which

---

2   The expression "approximately on your birthday" comes from the fact that leap years introduce a bit of a lag in the sequence of years, and it may happen that our "birthday" moment in a given year even falls on a date adjacent to our actual birthday. The problem is related to the question of when in a given year an astrological sign begins. More on this further in the text.

the other calculations of the positions of the planets and other significant points are based, whereas for the solar return calculation we know the position of the Sun (since it is implicitly recorded in the natal horoscope), based on which we search for the time of the year, when this exact position will recur. Proceeding from this known information, we find/calculate information for the other planets and points, and work out a horoscope, called a "solar return horoscope".

Nowadays, no one does this calculation on his own anymore. Even a natal horoscope, which is an incomparably easier set of mathematical operations and can be worked out in about fifteen to twenty minutes, is too time-consuming to be don made by hand in the computer age, while a solar return horoscope is so much more complicated and time-consuming that it is almost unaffordable for astrological practice without computer support...

A NOT-SO-NEW TECHNIQUE

The time-consuming nature of calculating horoscopes is probably the main reason why this predictive technique has not become popular before the twentieth century and, in particular, the advent of computers. Nevertheless, this is not a new approach. Thus, the English astrologers Kirby and Stubbs mention the popularity of solar returns in the seventeenth century and, in particular, Johannes Kepler's "habit" of "calculating a horoscope for his birthday every year". The famous English astrologer William Lilly also used solar returns in some of his more detailed case studies, and especially in the case of making horoscopes for important people from royal and political circles. Thus, in his book *Christian Astrology*, he published not only a description of the technique, but also a method of calculation for each year, based on the method of Maginus. The technique itself, however, is supposed to be much older, dating back to Greek times (Kirby & Stubbs).

The most detailed study of the solar return technique up to his time was made by one of the greatest astrologers in history, and perhaps the best astrologer of his time, the Frenchman Jean-Baptiste Morin (1583-1656). His collected work, embracing 26 volumes full of instructions and very concrete guidance, has influenced practicing astrologers right up to the present day. Morin was a man of enviable academic achievements, having studied medicine, philosophy, mathematics and astrology, and is known for his public polemics with his contemporary, the philosopher Rene Descartes, as well as for his appointment as Royal Mathematician at the French Court, and for his witnessing of the birth of the 'Sun' King Louis XIV. He collected and translated the works of Arabic astrologers into Latin, the academic language of the time.

After the eleventh century there were other authors, scholars and prolific writers, who drew on Arabic sources for their astrological (as well as medical, astronomical, mathematical and other) work, but it was Morin who took up the task most systematically in astrology. His monumental work, the *Astrologia Gallica*, brought rationalization to astrology, and in Volume 23 a thorough explanation of the then perception of the solar return technique is found, which Morin called the "solar revolution".

Individual volumes of this work began to be translated at the end of the 19th century, especially into Spanish, French and German. The brilliance of Morin's mind and the methodical derivation of his astrological theory did not find its way into English language until the last decades of the 20th century.

Just as Morin meticulously tackled the individual segments of the science of astrology, inventorying and analyzing the planets, signs, house systems, aspects, Arabic points, dignities and much more, he also examined the historical sources on solar returns with the same care. He found that at the time of Ptolemy (2nd century) they had not yet been calculated, mainly because the Greeks did not yet know how to calculate precisely the moment, when the Sun returned to the same place it occupied at birth, and also because the technology of timekeeping was not yet sufficiently developed. It was not until the time of Paulus Alexandrinus, who lived in the 4th century, that it became possible to calculate time more accurately, and the first mention of the solar revolution comes from his writings. The Arabs, however, were allegedly more involved in this technique (according to the historian and translator of ancient texts James Herschel Holden), "but not before the 8th century".

Although Morin rejected much of the Arabic and medieval astrological knowledge, he recognized solar return as a "useful technique" and hypothesised that *"the return of a planet to its natal position restores its basic power and ... germinates the seed of its own influence on the person born by producing its own effects..."*. According to Morin, solar return is the **chart of man's rebirth for the coming year, or the annual review of the celestial influences on the individual**. (Anthony Louis, pp. 115-117).

WHAT ACTUALLY IS A "YEAR"

Solar return horoscope gives us a picture of one year in the life of the person for whom it was calculated, from one birthday (moment) to the next. This introduces a new meaning to the term 'year', which is normally used to refer to the time between the two beginnings of calendar years, i.e. between midnight, 0.00 a.m., on the first of

January in the first year, and the same time in the following year. However, the concept of 'year' is not so simple, but rather more complicated than one might think. The term 'year' encompasses several different concepts, each representing a specific period of time or number of days, depending on the need or purpose, such as *calendar year, Julian year, leap year, fiscal year, academic year, seasonal year, tropical year, indefinite year, Martian year*, and many others. According to one official definition, a year is "the time between two dates of the same name in the calendar". But before we complicate matters further, let us write down a professional and perhaps most accurate definition of what a 'year' actually is, namely:

A year is the term for any period of time, derived from the orbital period of the Earth's orbit (or that of any planet) around the Sun.

One of the lessons, humans have mastered through years of training and involvement in business, production and IT systems, is how to do the accounts. From time to time we stop and draw a line under our work, say a project, a task, a period or an entire career; on one side we put the achievements, the good indicators and feedback, say rewards, financial statements and praise, and on the other side all the things that we cannot be satisfied with, having brought difficulties, lessons and repetitions, maybe even failures or resignations. Drawing the line in this way brings us a realization of our achievements or things worth remembering. Some people make such recapitulations more often, others less. The New Year, i.e. 1st of January - with the associated Christmas and New Year holidays, which give us a chance to breathe a sigh of relief as the rest of the world also takes a break from its activities for at least a few days - is somehow the most frequently used time line, suitable for taking stock, and this is also the purpose of various recapitulations, selections of the most important personalities, media and other events, achievements and the like, from Nobel Prizes to private reflections on a period. And there are not few people, who take time for a similar exercise every year on their birthday. After all, it is the day of changing our age number. Perhaps the best term for the time between two birthdays of a person might be "personal year", but this is not applicable in the case of calculating a solar return horoscope for some other entity, such as a country, a society, etc.

The resulting annual horoscope has all the elements of the horoscope we are used to from astrological work, and can in principle be interpreted like any other horoscope. However, throughout the astrological literature one can find quotations to the effect that such a horoscope "cannot stand alone", that "such a chart has no validity" or even that "it does not work". These rather radical dismissals are even true to a certain extent, as discussed further in the book, but the collateral damage is that over time the solar return technique itself has come to be regarded as "not useful".

In reality this is a simple and valid astrological chart, fully equivalent to, for example, a horary or electional chart. These two are also calculated to represent a moment in time, and from them we interpret the quality of that moment, and then infer the question asked at that moment, or the event for which the electional chart was calculated. And in the same way that a horary chart can provide not only an answer to the question posed, but much more, for instance the persons and circumstances related to an event, and even the possible development of the situation, so too can a solar return horoscope relate the whole story of the year it represents.

But in this case a problem arises. If we happened to find two people having the same solar return in a given year, they would then be expected to share exactly the same fate. But this is not really the case, due to the connection with the birth horoscope. For if solar return horoscopes were observed independently, without taking into account the natal planetary positions, they would indeed be identical or nearly identical, but the planets play different roles and are rulers of different astrological houses... A natal chart is thus a necessary and inevitable prerequisite for the interpretation of a solar return horoscope.

A Solar return horoscope, therefore, is an astrological chart, bringing information on the kind of a year a person can expect, which depends mostly on the respective person - on his or her potentials (natal/birth horoscope), on the personal development level, on external factors (parents, environment, cultural, religious, physical peculiarities, etc.), and above all on his or her perception. The same solar return horoscope is also a logical continuation of the previous solar return horoscope, since the two charts meet at the very moment when the Sun reaches its natal position in the year; one solar return is over and another one begins.

WHEN EXACTLY IS OUR BIRTHDAY

In everyday life, we have no problem with this question, because the date is recorded on our birth certificate, ID card and many other documents; moreover, we have been blowing out candles on the birthday cake on that very day since we were little. Astronomically and astrologically, however, the matter is not so simple, because it soon turns out that had we wanted to be quite precise, we would sometimes be celebrating our birthday a day earlier - or even later!

The Sun travels a full 360° circle of the ecliptic in a year, taking 365 days to do so, i.e. it makes about one degree of circle a day, returning to our birthday position on roughly the same date of the year. But because of the difference between 360 and 365, corrections have to be made first every four years, then more precisely every hundred years, then more precisely at the millennium, and so on. This means that every four years one day is added to February, i.e. 29 February.

This "enforced" day results in the signs not starting at exactly the same moment as, say, a year before. So, because of leap years, the ascendant is constantly moving in great leaps around the zodiacal circle, and the MC, the Midheaven, moves accordingly. This means that in each new annual horoscope these two points are in completely different places from the previous one, thus offering plenty of new possibilities for the interpretation of a solar return horoscope.

To see how this translates into practice, let us have a look at the following table. It shows the Sun's ingresses into Aries, i.e. vernal equinox, during the first 16 years of this century.

Table A

2001	20. 3.	13:32	2005	20. 3.	12:34	2009	20. 3.	11:45	2013	20. 3.	11:03
2002	20. 3.	19:17	2006	20. 3.	18:27	2010	20. 3.	17:33	2014	20. 3.	16:58
2003	21. 3.	1:01	2007	21. 3.	0:09	2011	20. 3.	23:22	2015	20. 3.	22:46
2004	20. 3.	6:50	**2008**	20. 3.	5:49	**2012**	20. 3.	5:16	**2016**	20. 3.	4:31

The table is based on Neil Michelsen's Ephemeris for the 21st Century. The times are taken for the Greenwich longitude (prime meridian, 0°). For Slovenia one hour should be added, for Ljubljana precisely 58 minutes. (As this is standard time, no additional correction of one hour is necessary.) Leap years are marked in bold.

First, we can see that for three years in a row the Sun enters the sign of Aries about 6 hours later, and then about 18 hours earlier (one day minus the fourth part of 6 hours…). Then we can also see that in 2003 and 2007 the Sun still entered spring on 21 March, and in all other years on 20 March. A slightly closer look shows that in each **row of** the table, i.e. at four-year intervals, the Sun's entry is pushed back in time - in the first row, showing the years 2001, 2005, 2009 and 2013, these times are 13.32, 12.34, 11.45 and 11.03, which means that the Sun enters Aries at about the same time every fourth year, but slightly earlier, each time by a little less than an hour.

This table also shows that our birthday moment within a year does not necessarily fall on the same date as our birthday. While it is quite possible that, at least in the beginning (in this case in 2003 and 2007), this calculated time falls even on the day AFTER our official birthday, our actual "birthday" then moves backwards (in this case, in leap years from 2044 onwards, it will be 19 March…).

Precession

All this is due to precession, the phenomenon that also clarifies the often-asked question why in some newspaper horoscopes a particular sign starts on a different date than in some other horoscopes…

> Precession of the equinoxes = a slow movement of the equinoctial points along the ecliptic. The result of this movement is that each year the equinox occurs slightly earlier than the end of the sidereal year. This results in the difference between the tropical year (365.2422) and the sidereal year (365.2563 revolutions of the Earth), which is tied to the seasons (and where the sign of Aries is always tied to the beginning of

spring), whereby the signs of the zodiac have long since ceased to occupy the same places they had occupied when calculated. Thus, over 72 years, the equinox point moves by an average of 1 degree, and over the course of one astrological era (e.g. the era of Pisces, which is about 2160 years long) by one whole astrological sign.

Precession, a special way of Earth's rotation, brings slow shifting of the positions of heavenly bodies and points, which – viewed from the Earth – move backwards. It is due to the fact that the Earth is not a perfect sphere, its center of gravity deviating from its center. The rotational axis of the Earth describes a conical shape, the earth moving like a spinner, wobbling away from the starting point. Consequently the vernal equinox point moves from East to West, i.e. in the direction contrary to the zodiacal circle. The period of its making a complete circle on the celestial sphere is called platonic year, numbering about 25.765 years.

Although the discovery of this astronomical phenomenon is attributed to the Greek scholar Hipparchus (between 146 and 130 BC), many sources testify that this anomaly in the Earth's motion was already known to some of his predecessors (in Babylon and, separately, in India) some two centuries earlier, although Hipparchus was probably unaware of their work. At that time the individual segments of the sky that we know as the "signs of the zodiac" were also identified, even though they sometimes differ in name in different traditions.

Precession is used consistently in Vedic astrology. The positions of the same observed planet in both types of astrology/calculation differ by about 24°, and the offset is called *ayanamsa*.

Many astrologers use precession in their calculations of predictive charts, in which case the calculated charts deviate significantly, with the difference logically increasing with a person's age. The relationships between the planets (aspects) remain the same, which means that, in the case of a solar return, much of the information an astrologer can offer to a client, remains very similar or even the same. However, the rest of the interpretation - not just the calculated points, e.g. houses or positions in houses, house cusps - differs to such an extent that the two principles cannot be exchanged or identified.

PRECESSION – YES OR NO

There is no shortage of advocates for the use of either zodiac, but the dilemma remains alive and thriving. If we narrow our discussion to solar returns, some authors strongly argue that a solar return, calculated without precession, has no value (e.g. Eshelman), while others use the opposite option, arguing that a solar return calculated in this way is also a useful technique. Some have studied the problem much more closely and have used both options. For example, the well-known astrologer Marc Penfield (according to A. Louis), "to test a hypothesis" for some time calculated all horoscopes in both versions. In his book, Louis himself illustrated this approach with the examples of the former Pope Ratzinger and the painter Dali (Louis, pp. 49-59).

In my own long-standing work with solar returns I do not use precession. The very essence of the principle, described in the present book, clarifies the reason: even if the planets in a precessed chart were to remain in the same houses as in an unprecessed chart, their positions would be altered to such an extent that the principle of intra-annual timing, elaborated in this book, would be useless.

This is not to say that the use of precession is a mistake or a step into the void, quite the opposite. There are many views, but the dilemma is much the same as in choosing a system of astrological houses - we cannot say that Koch works, and Placidus, Campanus and many others do not, or vice versa. Moreover, I occasionally exchange experiences with a good colleague, who consistently uses the precession calculation (according to Eshelman), and then we talk about *"solar return your way"* or *"fifth house my way"* and so on. And it's interesting to see how incredibly complementary information is...

At this point, however, it is necessary to make a digression. As evident in the further course of explanation, precession cannot be used with the solar return technique I am presenting - the two example horoscopes bring a lag, which undermines the very essence of my unique approach to this predictive technique, the timing, which is described in detail in Chapter 5.

Before that, in the next chapter, we will learn how to combine and use the natal and solar horoscopes to get the most useful picture of what the client or some other entity will be experiencing in a given year. Even if we use only the basic twelve elements of a horoscope, i.e. the ten planets and the lunar nodes, plus the twelve house cusps, it quickly becomes clear that the two charts offer a wealth of information. We just need to know how to use them.

But first we need to understand how to get to the solar return horoscope in the first place.

CALCULATING THE SOLAR RETURN HOROSCOPE

If not before, the moment there is something to calculate, astrology in our perception moves away from the pleasant, mystical and idyllic picture we initially got when reading astrology books or browsing astrology websites. Based on mathematical calculations of movements, it suddenly requires a lot of calculating as well as understanding of this infinite movement. Fortunately, today all this patient and time-consuming work is done by our indispensable computers, but it is still good to know the logic of how things work.

In the previous chapters it is stated that a solar return is a horoscope made for the exact moment in the year, when the Sun reaches the exact same longitude it occupies in a person's birth horoscope. The Sun travels all 360 degrees of the zodiacal circle in a year, which means that at any given moment it can occupy any position within it. Therefore, if we know the position of the Sun in the natal horoscope, we can find the same position within any year by a special calculation. This position in turn is the basis for calculating the solar return horoscope.

The accuracy of the Sun's position is of course crucial for the calculation of the solar return horoscope, since we know that it makes a 1-minute arc in time every 24 minutes; the accuracy of its position is also the basis for the calculation of the Ascendant and the house cusps that are the fastest to move in the horoscope.

For the same reason it is not possible to calculate and produce a solar return horoscope without a known birth time. In one day the Sun travels (approximately) a whole degree of the path, which practically means that solar Ascendant can be in any sign.

When we get down to it, it soon becomes clear why, until the advent of the computer age, solar returns were so rarely made. If we set about calculating our natal horoscope by hand, with a little practice and good concentration it can be calculated and drawn in a quarter of an hour, perhaps more.

Basically, we are looking for the positions of the luminaries and planets within birthday, which is a period of twenty-four hours, bounded by two midnight (or noon, depending on the type of ephemeris we have) sets of planetary positions, recorded in the ephemeris, namely midnight (0.00) of the day on which the person is born and midnight of the first day following. Simply put: if a person is born at exactly 6pm, this is exactly ¾ of that day, counting of course from midnight (0.00) to the next midnight (24.00) of that day. If we find in the ephemeris the two positions of the Sun at the two midnights, we can calculate the difference between these positions, or the distance travelled by the Sun on that day, and calculate ¾ of that distance. The same procedure is then repeated with the Moon and all the planets. Then, following a special procedure, we find the Ascendant and all the house cusps in the tables of houses, and the horoscope is made.

This is a rather simplistic picture, if only because few people are born at such a conveniently "round" time. With a birth time of 15.37, things get a little more complicated, and so does the calculation.

Calculating a solar return chart is essentially the reverse process, because we need to find the exact time within the day when the Sun will reach exactly the same position in the sky as at the moment of birth. We are therefore not looking for the position of the Sun, but for the time of day when the Sun will be at that exact position. This calculation is a little more complicated than calculating a birth horoscope, and it is also more time-consuming. At best we can manage to work it out in less than half an hour. But here I present a simplified procedure (for those who, with all the computer technology, would find such a calculation challenging):

- A day is 1440 minutes long
- ... Which equals 86,400 seconds
- 15.37 is 937 minutes or 56,220 seconds from midnight 0.00 (the start of the day)
- The quotient (percentage per day) in the first case is 0,6506944
- This quotient is then used to multiply the difference between the positions of a planet at the beginning and the end of the day (midnight positions).
- But before we get started, we should bear in mind that the ephemerides refer to the longitude of Greenwich (an observatory in the UK), which means that we also need to take into account the time difference between the longitude of the respective place and Greenwich, the prime meridian. This means that the position for midnight, as recorded in the ephemeris, does not apply to Ljubljana, as there is a difference of about one hour, which has to be added.

Example calculation

With an example of a solar return of a person, born on 1 July 1979 at 15.37 in Ljubljana for the year 2012, we are looking for the Sun's position between 1 and 2 July 2012 (the positions for Greenwich as written in the ephemeris):
- Sun position in the natal horoscope: 9°10'10" Cancer
- 30 June 2012 at midnight 8°37'09", 1 July, 2012 at midnight 9°34'20' Cancer (data: *Astrodienst Swiss Ephemeris 2012*, online)

- Difference between the two positions (= Sun's path on this day): 0°57'11"
- Converted to decimal: 57,183' (= daily distance)
- The distance the Sun has to travel to reach its position on the person's birthday in 2012: 33'01" (9°10'10" 8°37'09")
- Converted to decimal: 33,017'
- We use a simple formula:

$$\frac{33.017' \times 24h}{57.183'}$$

The result is 13.8574 hours, converted to a more accurate time: 13 h 51 min 26 sec
- Now we have to take into account the time difference between Greenwich and Ljubljana 1 h 58 min in summer (58 min in winter).
- 13 h 51 min 26 sec + 1 h 58 min = 15 h 49 min 26 sec
- We will therefore calculate the solar return horoscope for 1 July 2012 at 15:49.26

(A computer calculation test shows a very small difference, fully explained by the rounding of the values in the calculations, with a result of 15:51.46.)

SOME MORE MATHS AND GYMNASTICS...

In the same way, we calculate the positions of the Moon and the other planets, and then, with the help of the charts and the routine of calculating a personal horoscope, we search for the positions of the Ascendant and the houses, so that we can work out our yearly horoscope.

The accuracy of this calculation is most important for the Moon, which travels an average of about 12.5 degrees in a day, compared to much slower movement of the other planets. For a small relief, let us say that the slowest planets travel so little in a day, that we can safely write down an estimate between two midnight positions as their "position at 15.37". (Why just an "estimate"? Take the example of the movement of e.g. Uranus in 24 hours: from "19:00.2" to "19:03.7" (the way planetary positions are recorded in the most widely used Michelsen ephemeris today). The difference in a whole day is only 3 and a half arc minutes...)

Merriman's Cross Calculation

Different authors use different approaches to the calculation, which differ more or less only in the degree of simplicity with which they are presented. Perhaps the simplest way is that of Raymond Merriman, who, in his remarkable book, put the whole operation into a relatively simple cross-calculation:

$$\frac{a}{b} = \frac{c}{d}$$

whereby:

a - number of minutes difference between midnight and Greenwich time of birth
   (in our case "15.37" is 15 hours (= 900 minutes) + 37 minutes – 58 minutes difference between Ljubljana and Greenwich (or 1 hour 58 minutes, i.e. 118 minutes in summer time), which is 879 and 819 minutes respectively in summer time)
b - 1440 (number of minutes in a day)
c - the Sun's path from midnight to the time of birth (in degrees, minutes and seconds); and
d - the path of the Sun on that day (between 0.00 and 24.00)

Since we are looking for "c" in this procedure, the formula is inverted to the form

$$c = \frac{a \times d}{b}$$

The result is given in fractional seconds, which are converted to fractional minutes (the remainder are seconds). Add this to the Sun's midnight position to get the exact position of the Sun at the time of birth. Of course, here again we have to take into account the difference between the place of birth and Greenwich (Merriman, pp. 22-35).

Minutes/seconds to decimal conversion table (and back)

If we want to get the most accurate results when converting values, logarithmic tables are the best way to do it, but these are far from common use nowadays, and many people are no longer able to use them. Therefore, we can use a handy table to find a value either in minutes or in seconds, or in decimal form if we want to get the value in minutes or seconds from it. It is only important to be sufficiently focused on the calculations and not to confuse minutes with seconds or degrees...

Table B

Min/sec	Decimals	Min/sec	Decimals	Min/sec	Decimals
1	0,017	21	0,35	41	0,683
2	0,033	22	0,367	42	0,7
3	0,5	23	0,383	43	0,717
4	0,067	24	0,4	44	0,733
5	0,083	25	0,417	45	0,75
6	0,1	26	0,433	46	0,767

Min/sec	Decimals	Min/sec	Decimals	Min/sec	Decimals
7	0,117	27	0,45	47	0,783
8	0,133	28	0,467	48	0,8
9	0,15	29	0,483	49	0,817
10	0,167	30	0,5	50	0,833
11	0,183	31	0,517	51	0,85
12	0,2	32	0,533	52	0,867
13	0,217	33	0,55	53	0,883
14	0,233	34	0,567	54	0,9
15	0,25	35	0,583	55	0,917
16	0,267	36	0,6	56	0,933
17	0,283	37	0,617	57	0,95
18	0,3	38	0,633	58	0,967
19	0,317	39	0,65	59	0,983
20	0,333	40	0,667	60	1

The table is taken from Kirby & Stubbs, *"Interpreting Solar & Lunar returns"* (*"Tumačenje solarnog i lunarnog povratka"*, Metaphysica, Belgrade, 2006), p. 125, and was also used in the above example calculation.

OTHER ROUTES AND OTHER CALCULATIONS

The algorithm for calculating the horoscope of the lunar return and the returns of the other planets is based on the same principle, except that in these cases we take into account other time demarcations, i.e. the lunar month for the Moon or the periods of the individual planets. The resulting horoscopes will also accordingly be valid for the period between two returns of the Moon, or of a planet to a natal position, which takes about two years for Mars, about twenty-nine and a half years for Saturn, and so on.

There are, of course, more calculation options. Some schools favour the logarithm method, which is probably the most accurate, but what when the very word 'logarithm' makes us feel uncomfortable - few have seen it as something really useful and user-friendly. The reader will therefore not be bothered with details.

I have also attached a table with the daily orbital data and some other handy information about the planets' motions:

Table C

luminary or planet	average	maximum	minimum
Moon	13° 10' 35"	16° 30' 00"	11° 45' 36"
Mercury	1° 23' 00"	2° 25' 00"	- 1° 30' 00"
Venus	1° 12' 00"	1° 22' 00"	- 0° 41' 12"
Sun	59' 08"	1° 03' 00"	57' 10"
Mars	31' 27"	52' 00"	- 26' 12"
Jupiter	4' 59"	15' 40"	- 8' 50"
Saturn	2' 01"	8' 48"	- 5' 30"
Uranus	42"	4' 00"	- 2' 40"
Neptune	24"	2' 25"	- 1' 45"
Pluto	15"	2' 30"	- 1' 48"

Table D

planet	SU	MO	ME	VE	MA	JU	SA	UR	NO	EN
orbital time		1/13	0,24	0,615	1,881	11,86	29,46	84,01	164,0	248,1
daily travel	0,59	13,23	0,59	0,59	0,31	0,05	0,03	0,02	0,01	0,005
direct days	-	-	93	542	707	278	239	219	210	209
retrograde	-	-	23	42	73	121	139	151	158	157

III
STORY ABOUT A YEAR

So far we have been introduced to the astrological technique of solar returns, how to calculate them and how to prepare a combined chart in several ways, so as to interpret, what the person in question will experience in the course of the year between two birthdays. But what information can be provided by this chart?

Perhaps the best short answer is that the solar return relates the story of a year of our life. Just as a person's horoscope or natal chart gives us a glimpse of the whole story of our life, the solar return does the same with one segment of that life, limited to the time between two birthdays.

A solar return chart, therefore, is about what will unfold in a given year. It shows opportunities, challenges, interests and themes that relate to many spheres of life. To find out all this, the Sun, Moon, planets, Ascendant and other highlights, elements, groupings and configurations will tell us the story of the yearly horoscope.

Like any natal horoscope, a solar horoscope has all the elements that an astrologer uses in his work, so technically it can be interpreted in the same way. However, there is a fundamental difference, because this horoscope is not about the qualities and potentials of personality; rather, it is about the qualities of the period that the chart refers to, as well as about the potentials of our natal horoscope, which are supposed to be expressed in this very year.

We may learn that it is (or 'is going to be') a difficult year, a harmonious year, a relaxed year, a 'year of tied hands', and so on. Also, the highlights of a calculated solar return will indicate finances, health, schooling, moving, socializing... - all those areas that are of interest to a person, and at the same time the same segments of life that astrologers around the world deal with all the time and over and over again in the work with clients.

In general, a solar return horoscope is a unique forecasting technique that allows us to interpret the year for which the calculation is made, in the form of a story, limited to a specific period within that year or to the feel of an entire year; in the form of a series of annual horoscopes, it also shows the trend of the development of the person in question. It allows for many different insights or connections, contrary to other prediction techniques. In particular, it allows us to use the timing method, which I have developed, from a basis given to me back in the early 1990s, into an extremely useful tool, as discussed in the fifth part of the book.

MAIN ACTORS - TRAVELLERS - PLANETS

As in any horoscope, the planets also play a part in the story. The term "planet" comes from the Greek word for a "traveller, wanderer", because astronomers realized very early on that some of the spots in the sky move differently from others. There are eight planets (of which ancient scholars knew five, from Mercury to Saturn), including the Sun and the Moon. These two are not planets technically, but a star and a satellite of the Earth that reflects the sun's light, and are more correctly called 'luminaries', although they are usually called 'planets' anyway for ease of interpretation.

If we know how to interpret planets in a natal chart, we will have little difficulty in dealing with solar return, although we will find a few small differences as we go through the book. A tense Mars in such a year will bring challenges, quarrels, tensions, perhaps unhappiness or inflammation, depending on how it is aspected in this chart; Jupiter will "bring us luck" and open doors, while Saturn will bring setbacks and lessons. But we will learn much more about all this later. For the moment, let us just say that many of elements of solar horoscope will not allow for quite the same interpretation we use in interpreting natal horoscope, but this too becomes logical if we understand the principle of the time limitation of horoscope, in this case one year, and above all if we understand man's perception of himself and related events. We will see more about this, including the individual elements of a solar return horoscope, in Chapter 4.

TWO CHARTS OF THE SAME YEAR

After the calculation we get two charts, a natal and an annual one. We can immediately see they have nothing in common apart from the Sun's position. In rare cases, the Ascendants, and therefore the arrangement of the houses, are also quite similar. If we study solar returns a little more closely, we can observe the complex algorithm of how the Ascendants follow each other, with the Ascendant sign naturally repeating itself from time to time, but only in the 29th birthday chart it is almost identical to the natal chart. For the most part, however, the solar return horoscope is a completely different chart. In particular, the orientation is different each time and not even the same for two consecutive years - the Sun and at least the two closest planets, Mercury and Venus, which are never far from the Sun's position, are at the other end of the chart each subsequent year, and hence also the main focuses of the year, for which the annual astrological chart is made.

If we are to consider "what the year ahead will bring", it is not enough to only examine the annual chart. Moreover, such an approach never leads in the right direction. If interpretation were only restricted to annual potentials, in one year in the life of a very peaceful person we might 'discover' that she is going to become a terrorist, or that she is going to travel abroad, which is quite possible in some cases, but certainly not in the case of a single mother with three children. It is therefore essential to know, observe and study both charts, the natal and the annual one, and only on this basis to try to determine what highlights, developments and deviations the year in question will bring to the respective person.

At this point, it is necessary to emphasize the most important astrological rule of prediction, namely that all the potentials that are supposed to happen to a person are implied in his natal chart. This means that the single mother will not become a terrorist in that particular year, but that we will have to extract from her natal chart the more relevant potentials, related to the planetary positions, observed in the solar return horoscope. For if Pluto signifies 'terrorism' in astrology, we can at most speak of 'the expression (or activation) of this potential', i.e. Pluto, in a particular case or period within the respective year; the terms 'fear', 'powerlessness', 'pressure', a higher force, maybe even 'a nervous breakdown' or 'thinking about suicide' are probably more relevant in this case than 'terrorism'. The same applies, of course, to all planets and points of significance.

THREE VIEWS ON THE SAME TOPIC

From this point onwards we need to decide which direction to take. There are several paths leading from this junction, but we will only mention three, since there are mainly three principles of combining the natal and yearly astrological charts to be found in the world literature.

The oldest treats the two charts separately, comparing and combining them. The logic behind this approach is that until "recently" (just a few decades ago) astrological charts had to be calculated and drawn out by hand. This is also the main reason why this technique never became popular, and why history records only a few cases of astrologers who are known to have used it frequently or even regularly, like Johannes Kepler, who calculated the annual solar returns for his birthdays, or William Lilly, who was even able to use solar return horoscopes to solve some of his famous horary charts, if such a calculation could help him clarify the case (source: Kirby & Stubbs).

But even if we do not consider the solar return in isolation or on its own, it still relates the primary story of the year, and must be considered again within the framework of the natal premise. To compare such a pair of astrological charts means studying the yearly chart and checking in the natal chart, whether a potential, which in the solar return appears to be distinct, prominent, important or crucial, is likely to unfold, whether it is consistent with the potentials of the natal premise, or, more often, how a clearly expressed potential in the yearly chart, which we know how to interpret, would play out in the life of a person, who has more or less different potentials.

For ease of comparison solar return positions can also be inserted in the natal chart itself, which brings us to the second principle of dealing with the yearly horoscope. Let us say right away that this is the most common one of the three presented; it is done by inscribing the planetary positions, calculated for the solar return horoscope, on the appropriate places around the natal chart, thus obtaining two rings or sets of planets, including points like the Ascendant, the Midheaven and so on.

In this way we get an insight into the areas of life with active solar return horoscope potentials, which in turn will give us a lot of information to analyze and interpret the year in question. Particularly noteworthy are the lunar and ascendant positions, which give focus to the combined chart...

The third principle, presented in more detail in this book, is the inverse one. It also combines two astrological charts, the natal and the yearly one, but with the latter as the base with the planets and elements of the natal chart placed around it.

Natal and yearly horoscope

Natal horoscope shows our natal premise, outlook, potential for personal and spiritual development within a given incarnation. This potential is expressed throughout life, being slowly changed and upgraded; in some cases (when a major personality change befalls us in life, often expressed through a strongly occupied or accentuated eighth house, the house of transformation or catharsis) a person's life path changes to such an extent that we can even say that from a certain point onwards, "he or she is no longer the same person we used to know". But since this is also a possibility, visible in the natal chart, we can safely claim that the birth horoscope, as a potential, accompanies us throughout our lives.

If we therefore use natal chart as the foundation of the combined solar return chart and draw the elements of the solar return around it, it is easy - the foundation, the natal chart, i.e. our birth potential, is right in front of our eyes for all the years, i.e. each of our individual years. Just as with transit chart - for whatever moment in life we calculate it, the natal horoscope is there.

A solar chart, by contrast, is a picture of a year, the time between two birthdays. It could be treated separately, as a stand-alone chart, similarly to a horary chart. The latter relates a story or answers a question. In the case of a solar chart, the principle is very similar: even if we were to imagine the annual horoscope as a kind of horary chart, calculated for the exact moment in the year when the Sun occupies the same position within the year as at the time of a person's birth, interpreting this chart (and of course asking the appropriate questions, according to the rules of horary astrology) would lead us to a kind of "story of this year". It contains twelve houses and a whole set of planets, plus the two nodes and some other points, i.e. enough elements to give a fairly accurate picture of the theme, i.e. the year it represents.

The position of the Sun will therefore indicate the most important area for us to develop in the respective year. Mars shows where our energy will be invested (or where there will be conflict), while Pluto will point out potential fear, and so on. We can imagine that for the client, the year in question could be difficult, decisive, successful, non-descript or just hard and struggling, the possibilities and combinations are endless.

The accents on individual houses show the most highlighted or activated spheres, as well as the areas of possible victories, struggles, progress. The aspects are also informative, in accordance with our astrological expertise.

INTERDEPENDENCE AND INTERCONNECTEDNESS

It is difficult to say how many astrologers use this or that method in their work, but it seems the number of those, who place the natal planets on the outer circumference of the solar horoscope, is slowly increasing. This is no doubt also due to modern computer possibilities, which, in addition to the extremely fast calculation and printing of astrological charts, offer many things once undreamed of. Today's astrological computer programmes make it relatively easy to observe and study dozens of charts, a luxury that the great astrologers of the past were deprived of.

One of the proponents of positioning the solar planets on the outer side of a double chart, James Eshelman, argues in his book that *"the solar return* is a *technique based on transits"*, as it represents two charts with one point of conjunction - the exact conjunction of the natal and transiting Sun on the natal day. In this way, the solar return horoscope represents a transit chart on or very close to the birthday itself, which Eshelman says is *"in secret pact with the natal planets for the next year"*.

Before examining the reasons why drawing or not drawing solar planets around natal chart, let us imagine that the coexistence of the two charts can also be considered as the astrological charts of the two persons for whom synastry, i.e. a comparison of birth potentials, is done, except that in this particular case, the Suns of the two persons are exactly in the same place, while all the other planets and points are not. It is the interdependence of the two charts that we are analyzing and evaluating in the synastry. In the case of a solar return it is different: the interdependence of the two charts is not equivalent, since it is not two entities, whose qualities we are observing, but an influence or superimposition. Thus, understanding of natal chart is essential for the interpretation of solar return chart, because it could easily happen that we would find two people with exactly the same position of the Sun, and therefore with the same solar chart, but perceiving the same year in a fundamentally different way, because their predispositions, i.e. their natal potentials, are fundamentally different.

Inverse interdependence is less pronounced. The natal potential is clear and definite and, at first sight, quite independent of the potentials of solar return horoscope, making this pair of astrological charts a purely unidirectional dependence. However, if the sequential solar return charts are taken as a whole series of horoscopes, which describe in a graphic way the life, experience, perception and even the development of the person in question, it turns out that the potentials found in some of the charts of this sequence influence or interact with the development of the basic, natal potential to a greater or lesser extent. The planetary combinations between the two charts will tell us how and to what extent the two charts are intertwined, or to what extent the natal chart and its potentials are activated (and thus developed) by the solar return horoscope. Even if these are merely indications of the time at which some of the natal potentials are manifested, activated, developed, or fulfilled, there is a type of involvement in this direction as well. Thus, Eshelman's above-mentioned assertion of a kind of "secret pact" is quite accurate.

In this way, natal chart indicates the potentials that are there all the time and that condition many a decision or action. Thus, it is the natal horoscope, as a kind of supervisor, which "makes sure" that within a given year we act in accordance with our natal premise.

SOLAR HOROSCOPE - REALLY JUST A TRANSIT CHART?

With the realization that solar return horoscopes build on the basic premise, i.e. the starting point for the development of the personality and one's life path, and that they describe the year of the person, whose birth horoscope should always serve as the basis of a successful interpretation, the reasons for drawing the solar elements on the outer side of the natal chart are more or less exhausted. For if we imagine a whole series of such horoscopes, it soon becomes clear that we get a set of astrological charts, which are to some extent similar to each other, or which build on one another in a way, very similar way to the way in which we observe and draw out the transiting planets on an ordinary horoscope.

The first thing to notice is that the Sun is always in the same place, because it cannot be otherwise. Thus, the Sun in the eighth house of the birth horoscope is there every year - in the eighth house, the house of crisis, death, transformation, catharsis, etc., regardless of the length of the person's life. In each new solar return chart we can see that Pluto is slowly moving forward around its circumference, exactly as if its annual transit positions were drawn around it; but since Pluto is not retrograde for more than half a year, its position is never moved further back than the previous year.

The other planets move in a similar way, but faster, because their orbital period is always shorter than that of Pluto. We find Jupiter's position about one sign ahead each year, while Mars appears at different ends of the chart each year. The Moon, which is the fastest of all the bodies, also has its own laws of motion – for three or, exceptionally, four years in a row it will be found in the same element, for example, in the water signs, in a different one every year. It will then travel through the fire signs for a couple of years, and so on.

With Mercury and Venus, however, this principle fails. Namely, their distance from the Sun is less than that of the Earth, so they can never be more than 28 and 45-47 degrees away from the Sun's position. In a series of solar return horoscopes this means that they will always be "rotating somewhere close", i.e. at the same end of the horoscope. Mercury can therefore not be found in more than three signs; if the position of the Sun is between 28° of one sign and 2° of the next sign, however, Mercury will never appear in any but these two. If we used the equal or whole sign system, this would automatically also apply to astrological houses - Mercury could appear in only one of the three adjacent houses, or even in only two adjacent astrological houses, in any given year's horoscope.

The same is true of Venus, except that its extreme distance from the Sun's position is greater, which theoretically extends the range between its positions to four signs (about 94° in total) or five houses of the zodiac (in the case of mid-latitudes).

To illustrate this, we can examine the horoscope of the prematurely deceased rock'n'roll pioneer Buddy Holly and some of his successive annual horoscopes, where the birth horoscope is represented by the inner ring, and the annual horoscopes for 1954, 55 and 56 are arranged concentrically around it, the latter on the circumference of the combined chart. The progression of the individual planets, especially the slow ones from Jupiter onwards, is clearly visible, as are the positions of the Moon, which in this case completes its sequence of abiding in the earth signs (Capricorn in 1954) and continues in the air signs (Gemini and Libra). The Lunar Node also moves retrograde in a regular way, by 19° each year...

Of particular interest in this combined chart are the positions of Venus and Mercury. While the latter is in a very similar position in all the three years, Venus occupies almost the entire spectrum available to her, 46° to one side in the first year of this sequence of annual horoscopes, and equally far to the other side in the third year. (Note: The sequence of Mercury's positions in these charts may falsely suggest to the observer a motion similar to the other planets, but in reality this is only the result of the retrograde periods in these years.)

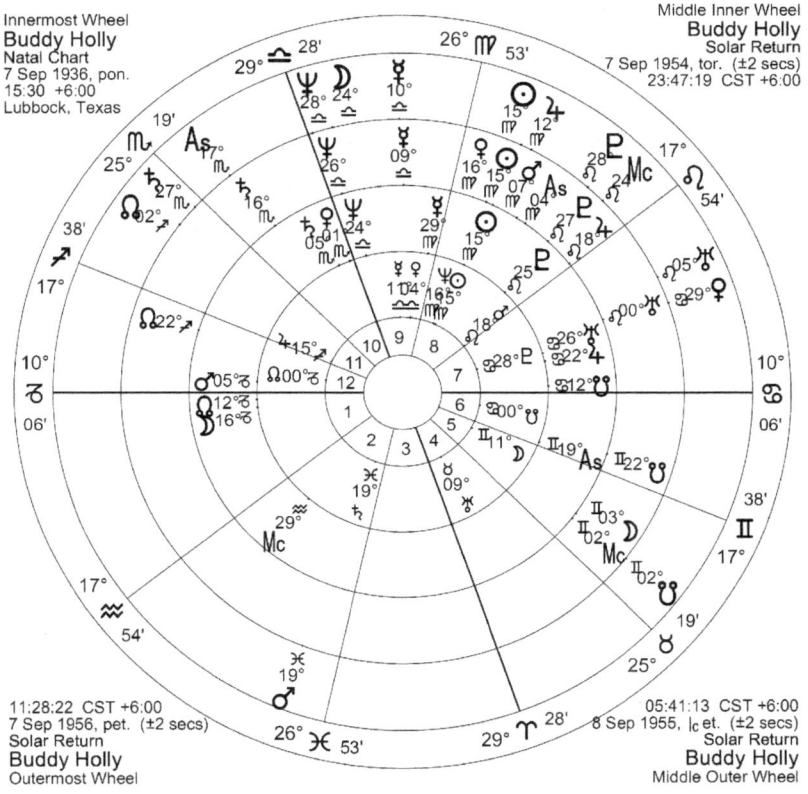

If we therefore had the possibility to make combined charts with more solar returns on the perimeter, we would get a good insight into the sequence of activation of the individual successive signs and/or houses, as discussed above.

TWO MINUS ONE

So, having learnt about the Solar Return chart as a transit chart, the dilemma arises as to why this particular chart would mean something more than any transit chart in a person's life. For if we calculate a transit chart for the next day, and the next, and so on, we get a series of charts, in which no planetary position changes significantly in the first few days except the Moon's, and the differences only begin to show up as time goes on.

If the solar planets are drawn around the natal chart, the solar return horoscopes are thus turned into a series of transit horoscopes - for just as transits of the planets proceed slowly along the ecliptic and the zodiac, so the accents in the horoscope change and build on each other, one after the other. In real life, this could mean that one year credit and investments will be highlighted, next year legal matters or education, then business, then friends or plans, clearly related to the eighth, ninth, tenth and eleventh houses. In real life, however, the highlighted areas of our lives rarely follow each other in that order...

As a footnote: no matter how long a series of solar return horoscopes we calculate, we will not be able to find certain positions in them, like Pluto in more than half of the astrological houses, and Venus and Mercury in the vast majority of them. The opposite is true if we interchange the two charts in a double circle, i.e. draw the planets and natal chart points around the solar return horoscope. In this way a whole new world of information and possibilities opens up.

The main advantage of such a combined chart is the insight into how different one year is from another. A series of horoscopes made in this way shows the sequence of years and their highlights; it is also possible to perceive individual patterns. The planets no longer move forward in the chart like transits, but occupy different positions every year, weaving a story of life and one's perception of events and turning points.

Combining the two horoscopes brings two sets of planets and other elements that are usually taken into account when interpreting an astrological chart. Each planet in this chart is natal and solar, so there are also two Ascendants, two pairs of lunar nodes, two Moons. In fact there are "two sets of planets minus one": the positions of the natal and the solar Sun are of course the same, that being the basis of this technique. We therefore suddenly have a much larger set to interpret the potential of a particular year in a person's life than before. Lots of possibilities for piecing together a story. And what is more – it rarely ever happens that some of these newly acquired elements don't fill in the unoccupied ("empty"[3]) segments, typical of most astrological charts.

THE DIFFERENT STORIES OF SOLAR AND THE NATAL PLANETS

It would be completely wrong to expect the planetary pairs in the solar return horoscope to relate the same or similar story. On the contrary, Mars on the outer side of the yearly horoscope (in our case, natal Mars) will act in a slightly different way than the 'inner', solar Mars. Let us therefore try to define the two horoscopes.

---

3 Somewhere in the slightly advanced stage of learning natal astrology, a student learns that the 'empty' parts of the natal chart represent "the possibility of development in the areas indicated by these unoccupied segments". This means that in life we should be given the opportunity to build up our natal potential even in those directions, where natal chart does not promise much fulfilment. This happens, for example, when the elements of the horoscope transit through or progress into those segments.
It is the solar return horoscopes that provide an excellent and tailor-made way to observe, in which years and how this will unfold. Thus, it very often happens that an astrological house, which is 'empty' in the natal horoscope, is fully occupied in a certain year. The spheres that are not of great concern to us or are less relevant in our lives, will at such times be activated. Moreover, the quality, intensity and coherence of the elements, that fall into such unoccupied parts of the natal horoscope in a particular year, show whether these are major, more 'fateful' shifts or developments, or 'just' attempts at activation or a foreshadowing of periods yet to come. In this part the solar return technique, especially the timing part, which follows in the next chapter of the book, can be significantly upgraded by experience, observation and statistics.

Years of experience with solar returns show basically two points of view. Roughly speaking, the solar, inner chart shows what is happening to us in the present year, our experience, our activities and their significance in our life; the outer, natal chart, on the contrary, shows us 'others', say other people or entities (e.g. a country, a flood, etc.), and also the circumstances that we are experiencing and are aware of, as well as the feelings that may result from various events, influences or perceptions, be they directly related to them or not. Natal Saturn or Pluto in the twelfth house of the yearly horoscope may thus indicate a stressed, depressed or frightened state, resulting from many things in the preceding period; usually it is not easy to ascertain the cause of such a state. Of course one of these planets may also indicate a clearly identifiable external pressure from a person, depending on the case. Solar Saturn or Pluto in the same house indicate a fairly clear condition the client is experiencing, as well as concrete problems.

In general, we can say that the natal planets inscribed on the outer circumference of the yearly horoscope indicate "something from the outside". In short, they are an indication of something that is "not ourselves", something that is brought about to us by others, or by life itself, something that we have no control over, but which we nevertheless experience as important enough to be marked in the yearly horoscope, and also something that we then react to.

Let us examine the case of Saturn in the second house. There, in the house of money, value and possessions, we like it even less than in other houses, because it means we do not have or will not have money. A solar Saturn in the inner ring of combined chart signifies lack of money. The exact events or circumstances generating this situation will be visible in the complete picture, but the money problem will be obvious. And unpleasant.

In interpreting this position, it is essential to be familiar with the person's natal chart in terms of money and value! This determines how serious or worrying the situation observed in the solar return horoscope may be. Only then do the connections within the solar chart etc. come into play.[4] However, if Saturn is found on the outer side of the natal chart, it is very likely that we will have the **feeling** of being short of money, without some financial possibility or income, but in reality, or in the end, things will not even be that bad. What is more, it may not even actually be the case afterwards. Is that even possible?

Will there be money or not?

The chances of this happening in modern times, when we experience many pressures and threats (astrology is always a reflection of the environment we live in), are many, and it is not uncommon to find ourselves in the position of running out of money, or being told there will be "no pay". Or a craftsman carries out work and issues an invoice, but the debtor is in trouble and is not able to pay. No doubt both cases will be recorded in the astrological chart as a threat and a feeling that something is going to go wrong, because as a result we may not be able to pay the bills or do something we had planned. This situation can last for weeks or months. Even if it

---

4   To illustrate: a strong and well aspected Saturn in natal chart will not bring so much pressure in a weak or strongly pressed position in the yearly horoscope, as a natal Saturn in a weak or adverse position. But the quality of the solar return second house itself is also important: if Saturn is found in the second house that begins in Sagittarius, and its ruler is well placed and strong, it will be quite different from Saturn in a "Scorpio" second house with a square to Mars, the ruler of the house, etc.

is 'just' a salary, with which the employer 'threatens' us two weeks before payday to 'motivate' or pressurize us, such fear quickly becomes serious. But then, on the due date or payday, the money does come, just as it was supposed to in the start. The 'threat' and fear eventually do not come true at all.

This example of "a craftsman and a debtor" is real and taken from the archives. The fact that the debtor would not have paid back the money by the agreed deadline might not have been acute, if the craftsman had not tied the money he expected to another investment on which much depended. The information that this money would not be forthcoming, as the debtor had no means available, was a serious shock. Where to go to resolve the situation? Fear and helplessness were clearly visible in the man's annual horoscope. But eventually the debtor got the money just a day or two before the deadline and deposited it in his account on time, thus solving the problem.

If the horoscope were perceived only as facts or accomplished situations, this problem would not even have turned up; after the battle, one can comment: "What's the problem, the money was paid as agreed!" But in the time **just before the payment that** was not the case at all, fear abounded! And if the horoscope shows our life, and therefore our perception of our life path, even if in a very small or specific segment or time frame, then 'the fear of money not coming' was also an important part of the story in that period. And such a feeling was indicated by Saturn on the outer side of the solar return horoscope.

The case of a person on bad terms with her brother and a job abroad

Examples to illustrate the principle are countless. But let us only examine two. They are real cases that have at some point found their way onto the astrologer's desk and are now part of a rich archive of cases.

In the solar return of a client, Mars, the ruler of the third house in her natal chart, and therefore the ruler of her brother, was on the outer side. A relationship with the brother, represented by Mars, cannot be good, and that was indeed the case. What is more, theirs was a really bad conflict. Even the loss of one of their parents did not change this, and the succession proceedings were carried out only through lawyers; the brother himself did not show up.

But for that year - without knowing all this at all - I predicted that the brother would come to talk to the client. One can imagine her astonishment, and also conviction, that this time I was really wrong. I gave the approximate month of the year when this would take place, and then we turned our attention to other things in the horoscope (especially as it was quite clear that this was a sore spot of hers that was unwise to touch...). However, the brother contacted her that very month, they went out for dinner and settled things from their past. (Mars, of course, does not represent dinner here, but the initiative - on the outer side of the horoscope, the initiative should have been his, however strange that may be - and this was indeed the case.)

SOLAR RETURNS

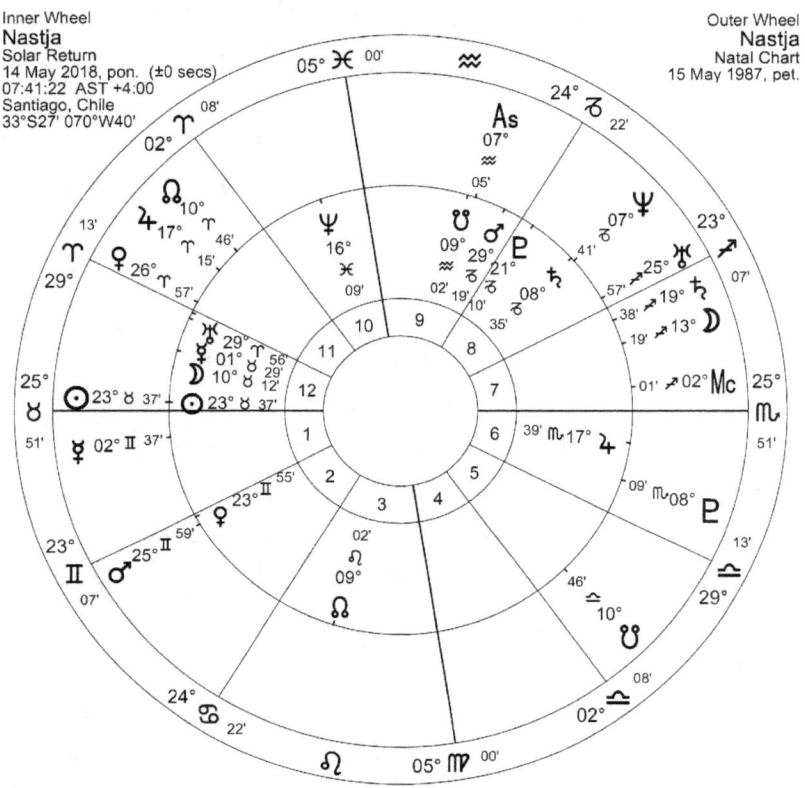

The second case is about the unfavourable situation of a woman working abroad and also about its outcome. The major highlights in the horoscope are a strong 12th house, the difficult financial houses, culminating Neptune in the 10th house and natal Pluto in the 6th house - everything to do with work and career. The background story is that she came to the country as a tourist, got a very good opportunity and a job, but later found another opportunity, which turned out to be anything but good. At the time she contacted me she was under enormous pressure and accusations, the owner of the business wanted to throw her out (natal Pluto) and threatened her with lawsuits (Mars in the ninth house), but - worst of all he had paid her nothing for several months (Venus strongly debilitated at the beginning of the second house). Solar Jupiter (ruler of the tenth house in this horoscope as well as in the natal!) pointed to a very favourable solution in the form of a second job, which was to come unexpectedly and also by a stroke of luck. As Jupiter also rules the eighth house in this horoscope, one could assume that this would somehow be related to money, and it was - she was offered a job by a man, whom she met by chance, chatting while waiting in line at the bank, and it turned out to be very good. Eventually her employer (the threat of natal Pluto) did not fire her at all, but she left on her own.

Natal and solar Pluto

An illustrative example of a record in the outer circle of the yearly horoscope is the development of events in the work and career of Miss J., which can be observed through two successive horoscopes and by means of feedback on events and intentions. Regarding the natal chart, this is a Sagittarius with the Moon in the 10th house, i.e. very sensitive to status and well-being at work, with a close conjunction of Pluto in Libra and Saturn in Scorpio (mutual reception of the planets). Both square Mars, ruler of the 7th house, signifying adversaries and enemies.

If the position of Jupiter at the cusp of the 10th house of the solar return horoscope may still show that all is well at work and that there is at most the prospect of more or better things to come, the square with Pluto in the 2nd house warns us that something is not quite right after all; the client indeed confirmed that in the area of salary on the one hand and a personal confirmation on the other, there were times, when she was no longer sure, whether this was still the job she wanted and would like to keep. However, the other planets in the 2nd house were bringing better times over the year (birthday: 5 December), even some small pay rise and a feeling that things might get better eventually. But when I reviewed the 10th house, I announced that things could get worse in July/August, underlining the possibility of pressure 'from above', perhaps from a change in management or even from external circumstances like a business review or inspection, certainly not something resulting from her inadequate work and the like. However, when looking at the next solar return, I mentioned the possibility of things escalating next spring, whereupon it was possible that she could decide (at least internally, if there was nothing that could actually be done) to leave, although it was not impossible that they would want to remove her from her post. Having been in no state to consider changing job, she simply took note of the information.

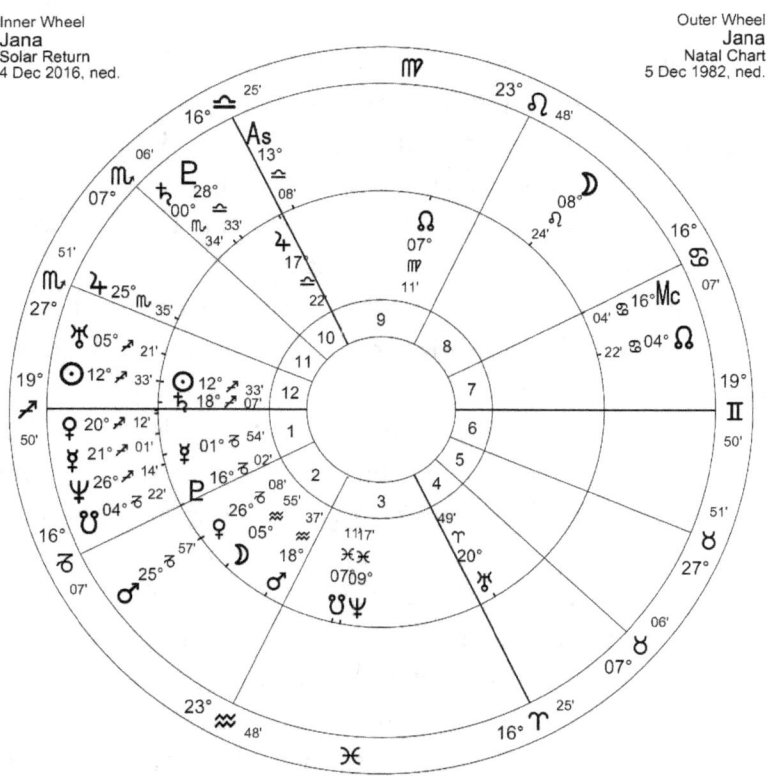

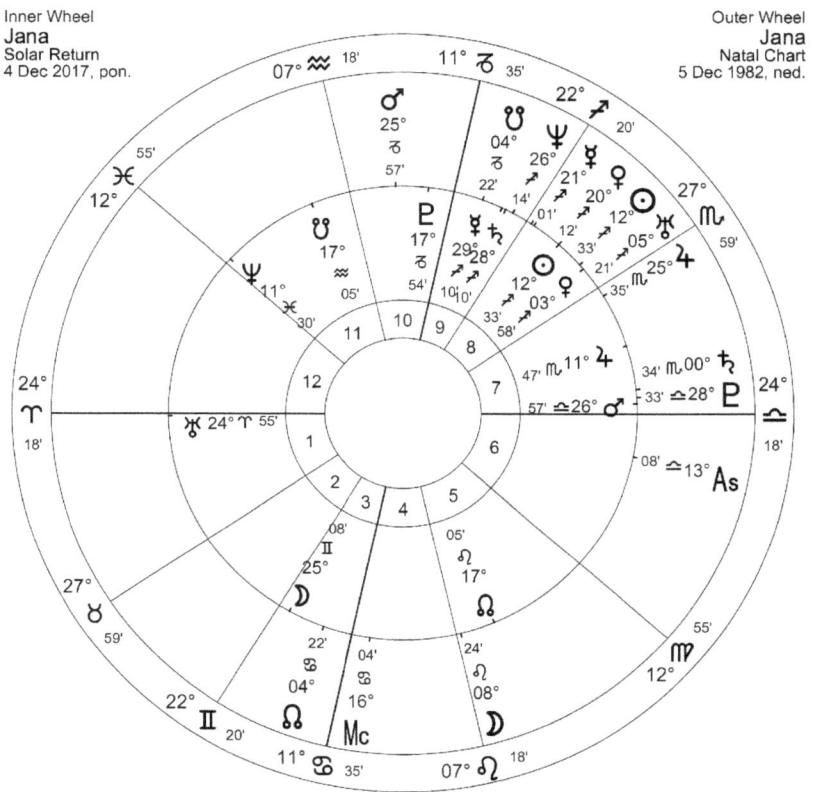

At her next visit, when both the periods had already been behind her, she said that "in July the pressure from the part of the boss really started to mount" (natal Pluto in the 10th house); he did not even want to hear that she was not only doing a good job, but was even performing better than most of her colleagues; in August, she was in fact subject to a lot of disgracing from his part (Saturn's position). A salary supplement just before her birthday (Mars at the end of the 2nd house in a trine with Jupiter in the 10th house) partly rectified matters, but nevertheless, sometime in the spring of the following year (2018, March or April) "the lid flew off the pressure cooker" and she firmly decided to quit the job as soon as possible. In astrology, the 'pressure cooker principle' is always signified by Pluto - at some point one cannot take it anymore! And in this horoscope Pluto is found exactly where it is supposed to be, if it is to illustrate such a feeling and decision-making - in the 10th house in the March/April 2018 term.

This time, it is a solar planet, not a natal planet, that shows action. The former two positions (Pluto and Saturn in the summer months) spoke of something coming from "outside" that she had no influence over, but Pluto's position in the second solar return chart shows her perception and her decision.

Miss J. managed to leave her job only after a few months and with considerable difficulties (natal Mars later in the year in the 10th house), even with the help of her personal physician (Mars is the ruler of the second part of the 6th house, the house of health, in the natal chart).

These examples are only a brief illustration of the possibilities of working with solar and natal planets, which will be discussed more closely in the next chapter.

IV
ELEMENTS OF THE SOLAR RETURN HOROSCOPE

ASCENDANTS OF THE ANNUAL HOROSCOPE

So far we have learnt that only by placing natal planets around solar return horoscope can we better explain why the successive years of our lives are so different from each other. One year we are doing great, the next year feels like someone has pulled the handbrake. If we know that the Sun is in the same place every year on our birthday, and that our personal planets Mercury and Venus - which, incidentally, are also rulers of the four signs, i.e. a third of the zodiac, and about the same number of astrological houses - are somewhere near it, as well as that the slow planets move only slowly forward on the ecliptic, rarely bringing any major shifts or differences, there must be "something" in the horoscope that accounts for, or causes, the great variations in the years.

But now that we know that the Ascendant can be in any sign on any given day, and at the same time understand the fluctuating cycle of the Sun entering the signs in successive years, which is due to leap years, it becomes obvious that the Ascendant in solar returns is in a different sign each year. This gives individual year/solar return a distinct tone and a completely different focus from the year before.

To a certain extent it is the Ascendant that gives a year its keynote. If for a natal chart the most personal points are the Ascendant and the Sun ("we are born in a certain sign"), this is not the case with solar returns, as the Sun is always at exactly the same place.

Every Ascendant has a unique set of astrological houses associated with it, which means that house cusps in each subsequent year are in different signs.

The Ascendant thus becomes a key factor in assessment of solar return and the associated story.

The order of the ascendants differs from the one we are used to in astrology and the zodiac, i.e. Aries, Taurus, etc. It has already been mentioned that every fourth year the Sun enters a sign at about the same time and on the same day, but with a slight delay, slowly moving backwards over the centuries. If this were not the case, we would only have four possible Ascendants in the solar return horoscopes, alternating within four-year periods. But all the Ascendants take turn throughout a person's life in a rather complex order, where the ascendant points are roughly repeated every four years, but at the same time they are constantly moving backwards around the zodiacal circle. This slowly reveals the pattern of how these ascendant points follow each other, so that we can roughly imagine in advance - for a few years - what the tones and highlights of these years will be.

A MARK OF THE ENTIRE YEAR

When working with clients, there can be a misunderstanding in terminology as to what is the current ascendant; when interpreting a yearly horoscope many people would ask "But now I'm no longer a Taurus ascendant?" Of course that is not the case at all, for the person with Ascendant in Cancer will continue to act in the manner associated with that sign, except that the annual ascendant will add a certain note to the annual horoscope. Our "Cancer Ascendant" will thus act in a more penetrative, aloof or confused way, according to the annual ascendant being in Scorpio, Leo or Pisces respectively. This is why it is necessary to be very clear in consultations with regard to the Ascendant we are discussing. Because of this ambiguity, I myself often use the term "mark" for the yearly ascendant, which is probably not the most appropriate term professionally, but is good enough to reduce the possibility of confusion when talking to people, who are not astrologically knowledgeable.

If we apply the term 'mark' in practice, we can say that Capricorn and Scorpio ascendants, for example, give a rather hard, difficult or unpleasant stamp to a solar return year, while Sagittarius, Gemini and Leo are expected to bring much more favourable years. This is just a basic rule, but let us look at these annual ascendants a little more closely.

FROM ARIES TO ... SAGITTARIUS

We start with **Aries**, which will undoubtedly give the year it marks an energetic, combative, perhaps even confrontational character. This will certainly not be a year of going with the flow; on the contrary, it will be a year in which we will want to impose our style, assert our will, make our mark, stand out. "The year of the Ram" is likely to be neither easy nor pleasant. It is also understandable that for many such a year can be a real nightmare (for example, for a Libra, who will have a hard time not being able to compromise and retreat, but will have to demand, be firm and decided and do many things a Libra does not enjoy...); However, for many others this will be a welcome period of progress, which may not have been possible before. The 'Year of Aries' is generally a time when we become more self-aware and perhaps stand up for ourselves. The position of Mars, ruler of the annual ascendant, understandably provides a more accurate interpretation guidance. Poorly placed (house, aspects) it will bring battle, strife, quarrels and hard work rather than self-assertion or even victory. But the house, in which this ruler is found, shows one of the most important areas of the year, a kind of a yearly focus.

As it is usual for neighbouring zodiac signs to be very different or even opposite in most of their characteristics, a **Taurus** year brings very different energies from an Aries year. It could be called a 'year of stabilization', of calming down, of taking a more sober view of the world and our problems. In such a year we try to keep things under control, to avoid big surprises and to stick to routine. For many people, finances will be of utmost importance, but, unfortunately, there is no guarantee for a Taurus year to be financially favourable. A closer look at the position of the annual ruler of the Ascendant, Venus, will of course be more revealing. But that is not all. As the sign of Taurus forms the so-called 'Series 2' with its ruler Venus, we associate them with the second astrological house, which is not only about finances, money and prosperity. However, since this is also the house of self-worth and one's level of self-confidence, the years with Taurus ascendant are the ones when the potential for progress in this highly sensitive area is very pronounced. The term 'stabilization' can also be understood in this sense.

Rather similar are the years with **Virgo** ascendant of the yearly horoscope. As a rule, such a year is not dramatic, quite the opposite. And yet it is a year in which, on the one hand, we strive to keep things under control, except that there will be a lot more analyzing than usual. On the other hand, we are at least partly aware that things are getting out of hand or are speeding up, and we are uncomfortable at the thought that now something will have to be done, a decision will have to be made, an initiative will have to be shown, a fight will have to be fought; not a good prospect for a conformist Virgo. We will seek information to help us make better decisions, we will seek opinions, but generally we will not be completely sure that we are right. A 'Virgo year' will bring a strong focus on health for many, and not necessarily on the health of the person in question. Virgo is caring, likes to help and take care of others, so it is quite possible for the year to be about the health of others. Another concept worth mentioning in reference to a Virgo ascendant is the need to bring order to life at all times. In one way or another. More information can be provided by the horoscope itself, especially by Mercury's position as ruler of the solar return's Ascendant.

As mentioned above, the annual ascendant follows a complex algorithm, but some of its fragments are quite common and recognizable, making a kind of "pattern". Since Virgo begins one of these fragments, let us look at it in the order of the signs of the Ascendant, which we often encounter.

Virgo is followed by **Scorpio** regarding the year marks. If a Virgo can still maintain some level of control, the year ahead will turn everything upside down. It will be a year of breakthrough, crisis, battle, intolerance, radicalism, certainly nothing very peaceful or tranquil. For many this is a year of personal transformation, at least partly, perhaps a year of breakthrough in some area, something deep and solid. For many, such processes are going on somewhere inside without any outward sign of a profound shift, and yet in such a year one will be forced to act. The positions of Mars and Pluto, the two rulers of Scorpio, will be more revealing, along with a thorough consideration of the eighth house and related matters.

A Scorpio year is followed by a year with the Ascendant in **Capricorn**. Generally speaking - and the term has been tried and tested in literally hundreds of cases - such a year can be called a 'year of tied hands', and many people have said it felt like having an iron ball strapped to their leg... We know what we have to do, we know what we want to achieve, and even what steps should be taken, but somehow nothing works. Something is standing in the way - tradition, powerlessness, circumstances, people, character traits, calculus; the time is definitely not right yet and changes will come when Saturn, the ruler of the current ascendant and also the master of time, will allow for them. This will be in the next year with ascendant in the sign of Gemini. The Capricorn ascendant year is also a year of reflecting upon status, job, career, possible changes, problems, even health, chronic illnesses, and anything else related to Capricorn, which Saturn's position in this year indicates even more precisely.

The Year of **Gemini** is generally a year of change, a year of action. After struggling to hold a position with information and - possibly painless - corrections (the 'Year of Virgo'), after being hit by an unwanted dress rehearsal in the 'Year of Scorpio', and after having to wait for change in the 'Year of Capricorn', change is finally here. Since Mercury, the ruler of Gemini, multiplies events and meanings, there will be no boredom, nor will there be any opportunities to get involved in just one area or activity of our preference. We are more likely to be caught up in different events that will alternate and intertwine, so things will be quite tiring at times. We will also "speak up" this year, discussing problems, consulting and finding solutions.

Significant life changes are possible, depending on the focus of the observed solar return, shown by the ruler Mercury, the Sun, planetary stelliums and more, as in all other cases. We may change job or workplace, buy a new car, improve communication with neighbours, change a life habit, relocate ... possibilities are many in such a year. There will almost certainly be no boredom...

The annual Ascendant in **Cancer** is not the best indicator of the year in question, unless of course you are supposed to be dealing with family, children and family business, including estate business, or if you deal with something retrospective, perhaps even build or renovate a house. Cancer mostly goes backwards, and a Cancer year is more likely to be a step or even several steps backwards, rather than progress. Tradition will be important, as well as piling things up, old matters, security issues. In general things tend to move more slowly, there is a lot of checking and procrastination. Safe routes and options will be sought. The best part of a Cancer mark year is an opportunity to do something that has been waiting for and should already have been done, so we will not have to deal with it in future. A year with Cancer ascendant will certainly give us every opportunity to do so. If we had to decide whether this was a "good" or a "bad" year, it would probably not be considered the best...

For a year with **Leo** ascendant, one would expect something great, brilliant, big, exquisite - and from Leo and the Sun, the ruler of this fiery sign that gives us vitality, that is what we mostly get... However, many years of studying and interpreting solar returns, and especially the feedback of people, who know more about these years because they have experienced them, shows that a Leo year, which, while bringing a lot of drama and hard work to make our way through the challenges we face, is certainly much more than "nice, smooth and perfect". The good thing is that we do not run out of energy to win our battles, and the position of the Sun in such a horoscope (one of the key elements for interpretation) shows us the potential crucibles as well as our chances of riding out such a year as successfully as possible. In general, this year will not be remembered for the good, except by those who have waited and toiled for recognition and success (fame). The latter may indeed arrive, but such an outcome must be very strongly supported by the rest of the horoscope.

A year with **Libra** ascendant promises exactly what Libra usually brings - balance. A working label for such a year might be 'a year of harmonization'. And since Libra is not the easiest sign to understand, offering us the widest possible spectrum of all feelings, we can say that even for such a year the possibilities are many. In terms of harmonization it can be about settling relationships or court cases, important agreements, team work, perhaps even public relations as a novelty in life; however, all too often this 'harmonization' is more about harmonizing within ourselves than harmonizing with others. This is the case when we achieve a key insight, take a personal growth course that will prove to be of long-term significance, and so on. Definitely something that will make a real mark on the whole year, otherwise it would not have been written in the yearly chart as the leitmotif. It is also possible to hesitate, especially when one is normally not inclined to hesitation, or to ask for another's opinion when one usually decides everything by oneself...

It often happens that a Libra year follows a year in which a person's life has undergone major, perhaps even remarkable changes that no longer allow them to live in the same way, in the same environment or atmosphere. In such a case, the two cups of our inner scales are so agitated and fluctuating that it takes time to calm down, the 'Year of Libra' being perfect for this. The term 'harmonization' makes perfect sense in such years. But it also means that the year is not necessarily peaceful - just the thought that it is 'significantly more peaceful than the previous one' is enough, and that is already a very relative concept ... just as our perceptions can be relative.

But we should also not overlook the possibility that this 'harmonization' may also mean the principle of 'an eye for an eye', which is also a Libran principle - especially in the case of a longer or deeper anomaly, which may be ironed out in a year of Libra. Of course, the position of Venus, the ruler of Libra, provides more information. So, having just won a lawsuit, we may now be countersuing to get even ... the possibilities are manifold.

If the sign of **Pisces** usually means "something that is and at the same time is not", leading us into a kind of vagueness and confusion, a year with this sign on the Ascendant will be no different - things will be happening that we have never imagined, but not exactly the ones that should have been taking place; we will be lost and searching for a compass, with plenty of inspiration, artistic flair, great feeling, even luck in getting out of difficulties, perfectly in line with Neptune. This will not be a year of great achievements, unless they be in line with the well-placed potentials of Neptune or Pisces in our personal horoscope - in which case success within a Pisces year can indeed be Success with a capital "S" ... As usually, Neptune will force us to investigate all possibilities, to separate the wheat from the chaff, and to decide for ourselves what is right and what is wrong, without needing outside information ... which, anyway, will confuse us along the way more often than not. And as in every case, here too - the positions of Neptune and Jupiter will be most relevant for the quality of the Pisces year. Footnote: in the 'Year of Pisces' our frequent commentary will be: "But that cannot be true..."

The contribution of **Aquarius** to a year will be different, just as everything to do with Uranus is different. The themes of the Aquarius year are independence, significant new developments, or a sudden breakthrough in quality, whether in business, work, finances or relationships. We will be more independent and able to say "I don't care" in the face of problems (even if we usually find it hard), we will take matters into our own hands and do things the way we have always believed should have been done ... regardless of tradition, rule or opinions of others. If that requires conflict or confrontation, it won't stop us - the idea of what needs to be done will be instantaneous and more important than the way to get there. This will be a year of sudden upheavals, of shocking job cuts (no, they won't fire us, we will fire them!), of starting new, completely different paths or businesses, of becoming independent from our parents (to whom we have not dared to say *No* for too long...), and so on. In fact, anything can happen in a year like this - except routine, predictability and boredom, quite in line with Uranus. The latter will reveal a lot more about the year ahead, however, we should also take into account the traditional ruler of Aquarius, Saturn.

In listing the ascendant signs of solar returns, **Sagittarius** is not the last in line by accident. For if there are no very unfavourable highlights in such a year, like a very negative position of Jupiter, then this will simply be 'the year of liberation/relief/expansion/ knowledge/progress'. Very often it follows a year that is just the opposite – gloomy, problematic, full of hard work, lessons and frustrations. So in a Sagittarius ascendant year one can feel relieved. *I can finally breathe!* or *Let's open the windows!* or *It's over!* - whatever it was -, are the frequent comments. Sagittarius years can of course also bring the potential for ignoring or neglecting important things, making a situation worse due to our own stupidity, or overlooking facts/information, but usually a 'Sagittarius year' is one of the more auspicious in the whole sequence of years; one only wishes there were more of them...

What if the Ascendant is placed at the end of the sign

As in the case of natal charts, Ascendant sometimes falls in the last degrees of a sign, so that the entire first house falls in the next sign. The logical question is, which sign to give priority to in determining the "mark" of the year. The principle is the same as in natal astrology - even if the ascendant is at 29°59' of a sign, it is placed there and not in the next sign, which is indeed the one immediately following. It is true, however, that the next sign, i.e. the one that fills the space of the first house in the yearly horoscope, will play a very important role and should not be underestimated in the interpretation. Moreover, it may well be that in such a year there will be more focus – although perhaps not initially - on the sign that follows...

It is also very often the case that such a year can be labelled with the basic characteristics of both signs, the one on the Ascendant and the one occupying the majority of the first house.
Let us for example take the Ascendant at the end of Gemini,

which is supposed to bring 'a year of change', with the first house mostly in Cancer; the label of such a year can no doubt be corrected to 'a year of change, but slower change, or change that will have to wait a little longer'. Or, for example, a year with Ascendant at the end of Leo: here 'a year of drama' will turn into 'a year of controlled drama, unfolding away from the public', perhaps even 'a year of dramatic health problems'. Or, last but not least, a year with Ascendant at the end of Capricorn: a 'year of tied hands' here turns into 'a year when we can't move forward, at least not for a while, but somewhere in the course of time things suddenly change, and so on.

Algorithm fluctuations

In the above text the common order of the Virgo – Scorpio – Capricorn – Gemini annual ascendants was mentioned. We can attribute it to the old folk expression "two bad years", whereby the adjective bad refers to the middle two signs, since Gemini already brings changes. In practice, however, it often happens that one of these ascendants falls at the beginning or at the end of a sign, say at the end of Scorpio. In this case the next year's ascendant will no longer fall into Capricorn at all, but already into Aquarius. The exception is most welcome, as we will not have to go through another 'year of tied hands' before events, changes and breakthroughs finally unfold (in this case not under the rulership of Mercury, but of Uranus), as they will do so sooner. Similarly, if the ascendant at the beginning of Capricorn is not followed by the ascendant in Gemini but Taurus - even if a 'year of change' is already 'on the cards', it will have to wait a little longer, or be more predictable, allowing us to prepare; changes might also be happening at a slower pace, Taurus being a slower sign...

There can, of course, be many such fluctuations. These variations make interpretation somewhat more complicated, but they also bring more variety and challenge to our lives - and to the astrologer's work.

„Swap" option

Even if we have chosen to put natal planets around solar chart, solar ascendant is a point that is also useful to be checked in the natal horoscope, which would require a reversed view, that is, one in which the birth horoscope is in the inner circle of the combined chart. The Ascendant as the 'I point' falls in a different sign each time, and thus in a different house of the natal horoscope, giving us additional information about the focus of that particular year's horoscope in the person's life.

To find this out, it is not necessary to cast an additional yearly horoscope; it is enough to search for a button in the astrology software that gives this option instantly, and after examining the chart, returns you just as quickly to further study of the solar chart. (In SolarFire software, this is the "Swap" option).

THE SUN - THE CENTRE OF ACTION (AS ALWAYS)

The principles for interpreting all types of returns are the same - we take the planet, for which the return was calculated, as a starting point. Everything revolves around it, this planet becomes the main actor in the drama, while the other planets and elements become side-actors, supporting the role of the main character. The Sun's position, with its aspects, will bring energy and a strong influence on the events of the entire year, so it should be interpreted first.

However, there are also exceptions and other indicators, making a solar return more than just an "interpretation of the Sun in a particular year".

The planets and their mutual relationships – the aspects – also play a part in the solar return story. A tense Mars in a solar return chart will bring challenges, quarrels, tensions, perhaps unhappiness or inflammation, depending on how it is aspected; Jupiter will "bring luck" and open doors, Saturn setbacks and lessons. But we will learn much more about all this later. For the moment, suffice it to say that many elements of solar return horoscope will not allow for quite the same interpretation we use in interpreting birth horoscope, but this too becomes logical if we understand the principle of the time limitation of solar return, in this case one year, and above all if we understand man's perception of himself and the related events.

One of the common questions at a certain stage of getting to know a person is that person's "horoscope sign". This term astrologically refers to the person's sun sign. This information is used in many aspects of identification, the most common being the national identification number in some countries, where the date of birth is used as a part of the code. Knowing our client's sun sign thus reveals a first piece of information, a partial identification of that person. Astrologers know very well that this is insufficient, however, this information is in most cases still more important than the information about Saturn, Pluto or Chiron. Moreover, it is with the Sun, which sets the tone of the whole natal chart and the person's character, that most astrologers begin their astrological analysis.

THE SUN IN THE SOLAR RETURN CHART

The Sun is the centre of our Universe, the centre of our system, the source of light, energy and vitality, life is impossible without it, and, without its radiation, the Earth would be a cold piece of rock lost in space like millions of others. . That is why in our natal chart the Sun represents the essence, authenticity, identification, revealing who and what **we are**. Its energy (and also its position in the chart) is our starting point. If we don't express and radiate its energy, or if we block it, we stagnate or slowly fade, we disappear, which can be expressed through a lack of vitality, a sense of fulfilment and contentment. We are neither well nor strong, perceiving everything as meaningless; there is also the danger of illness, slow denial and ultimately self-destruction. The Sun thus shows us the way to integrate the planetary energies on the way to realizing our life purpose and creating an integral personality. The aspects of the Sun in the natal chart show how other planetary energies are involved in this process.

Here we come to a small embarrassment. The position of the Sun in an astrological sign will not tell us anything in a particular year - Aries is Aries this year, and will be Aries the next year, and every year after! Our natal Sun is locked to the same point for life. The progressed Sun, however, will continually signal our evolution and at some point change signs, becoming our super sign, but the natal Sun remains the base and starting point for life. The transiting planets will add an on-going, incidental, sporadic additional information about the influences without changing the essence.

At this point we come to the essential advantage of placing solar planets in the inner, initial circle of the solar return horoscope. For if we take natal chart as the basis of the calculation and draw solar returns around it for each year, as described in the previous chapter, the Sun is not only in the same sign every year, but also in the same house. No evolution?

If we swap things, however, we get plenty of development. The Sun moves from year to year in some traceable and already described sequence. But the crucial thing is that we find it in a different house every year, which indicates a trend, an evolution.

The house, in which the Sun is 'found' in a particular year, will show the area of work, of seeking opportunities for fulfilment and progress, of strengthening our positions. This is the area that will represent a kind of focal point or red thread of all the events of the year, a field of our creativity, of the struggle for ego, of achieving recognition, a field of brilliance and noticeability. It is in this field that we will strive, excel, struggle and win - or perhaps not, if that is what the solar return says...

The Solar Return House, where the Sun is placed for a year, points to the area, where we are expected to make most progress, the area of the most important lessons of the year. If we try to avoid them, life itself - the evolution of the position in the segments of our lives, related to the position of the Sun in the year's horoscope - will confront us with challenges (most likely ones beyond our control) and force us to accept them and try to overcome them. Otherwise, i.e. if acting in accordance with the potentials of this period, we will develop a stronger sense of our own power and identity. The position of the Sun therefore points to the field of the strengthening of one's position and, consequently, to the deepening of the self-knowledge and purpose.

12 astrological houses

The solar return horoscope naturally contains the same houses as the natal horoscope and is therefore interpreted in the same way. They can be summarized as follows:

1st house: self, behaviour, vitality; 2nd house: (earned) money, goods, values; 3rd house: siblings, environment, relationships, communication; 4th house: home, security, roots; 5th house: children, love, creativity; 6th house: work, health, habits; 7th house: others, partners, marriage; 8th house: passions and crises, transformation, financial transactions; 9th house: foreign affairs, studies, higher knowledge; 10th house: profession, occupation, career, success, reputation, status; 11th house: friends, plans, joint projects, and 12th house: enemies, hidden affairs, problems and loneliness.

This will obviously not be enough. Even in natal astrology such a brief definition of each house is not sufficient, since there are only twelve spheres of life, which can be illustrated by only twelve compartments, into which we have to classify, according to some common characteristics or denominators, everything that we know, or more precisely: everything that can be astrologically qualified. Sometimes, however, there are special or strange matters of interest, such as a lost wallet, a sick cat, a neighbour's tree, pasta, a bad joke or a broken bicycle. It is sometimes incomprehensible what tourism and a professor, for example, or a dog and an aspirin, have in common (sharing the same drawer), but if we understand the essence of the individual astrological principles, we can arrive at relevant answers.

But there is more. Even if, for example, the eighth house is associated with transformation, it is clear, purely experientially, that there are two basic views here: transformation is the overall process of our life, because only in this way does our whole incarnation path make sense; and individual major or significant steps on the path of this transformation only take place here and there (astrologically determinable from several angles or by several predictive techniques), so in some years yes, but in others no. The traditional astrological meaning of the eighth house ("house of death") is even less useful in interpreting solar returns. In an expected lifetime, the Sun will be in the eighth house (whichever house it is in the natal chart) about eight times - we are certainly not likely to die that often... Moreover, it is statistically refutable that the death of an individual would be related to this particular position of the Sun in solar return horoscope. If we were to look for an indicator of death in solar returns, we would find it in many other houses and especially combinations...

It is also not enough to explain the meaning of the eighth house in solar returns only in terms of 'crises' or transformations; we need a wider range of concepts that come into play in our interpretation. The following table offers a slightly expanded list of possibilities.

Table E

House	Explanations	Natural ruler
1.	The first house symbolizes the rising of the sun; self, personality, physical body, vitality, character; head and face, upper jaw; appearance, mask, facade, how we present ourselves externally, self-expression, self-confidence, personal freedom, freedom of choice, pioneering, initiative, personal action	Mars
2.	money, wealth, earnings, salary, resources, earning capacity, material security, personal values, worth, possessions, movable property, ownership, practicality, talent; neck; self-consciousness, self-esteem, self-importance (especially in one's own eyes); throat, ears, lower jaw, neck	Venus

3.	siblings, relatives, neighbours, neighbourhood, short journeys, speech, communication, telephones, writing, learning, primary (and sometimes secondary) school, short journeys, road accidents, vehicles, press, rumours, ideas, (dealing with) lawyers; shoulders, arms, hands, fingers, lungs	Mercury
4.	end of anything, grave, foundation, soil, real estate, birthplace, homeland, home, roots, parents (especially father), security, family life, ancestors, past experiences, domestic affairs, tradition, house or dwelling, building, renovation of dwelling, savings; stomach, breasts, womb	Luna
5.	creative self-expression, offspring, children (especially first), pregnancy, birth, love, love affairs, courtship, romance, gifts, pleasure, entertainment, show business, hobbies, recreation, sports, play, gambling, betting, risks, love, sex, engagement, ornamentation, glamour, jewels, leadership, sports leadership; heart, circulatory system, nervous system	Sun
6.	work, job, colleagues, working conditions, tools, favours, service, servitude, apprenticeship, subordination (including subordination to the will of others), duties, habits, routines, details, living conditions, order, analysis, tidying, servicing, help, hygiene, health, pets (up to the size of a goat); abdomen, intestines, nervousness, psychosomatic illnesses	Mercury
7.	partners (private and business), „others", public, rivals, opponents, open enemies, fight, war, negotiation, diplomacy, balancing, talks, cooperation and business cooperation, public relations, teamwork, marriage (especially the first one), divorce; agreement, contract; kidneys, organs in pairs (the lower part of the body), veins, ovaries; sunset	Venus

8.	resources or money of others (e.g. a partner), joint money, (joint) investment, investment, shares, debts, taxes, inheritance, legacy, investigation, detective work, transformation, crisis, destruction, decomposition, collapse, poison, garbage, sewage, extremes, intense feelings, sex (in the sense of preserving the species), orgasm, sexual perversion, pressure, blood, bleeding, surgery, regeneration, fears, verge of death, near death experiences, death, nervous tension, loss, occult; bladder, prostate, haemorrhoids, anus	Pluto and Mars
9.	foreign countries, long journeys, tourism, far away, justice, law, court, lawsuits, legal and administrative matters, permits, documents, higher knowledge, (higher) education (studies and postgraduate studies), philosophy, search for truth, spiritual development and practices, belief system, religion, intellectual activities, politics, science; thighs, hips, varicose veins	Jupiter
10.	honour, reputation, fame, success, career, profession, job, bosses, employer, superiors, authority, pressure from above, responsibility, ambition, social status, leadership, governing structures, judge, desire for authority, parent (usually mother), promotion, rewards; knees, bones, teeth, skin	Saturn
11.	circle of friends, associates, supporters, associations and acquaintances, club and group activities, social connections, aspirations, desires, perspective, tomorrow, future, planning, charity, (charitable) organisations, entrepreneurship, independence, networking, social awareness; lower legs, nervous system, spine, joints, arteries	Uranus and Saturn
12.	illness, grief, closed spaces, closed institutions, loneliness, hospitalisation, prisons, hospitals, mental hospitals, convents, research, working behind the scenes, conspiracy, terrorism, hidden enemies, undermining, secret societies, deception, self-containment, depression, nervous breakdown, powerlessness, dissatisfaction, chronic problems, suicidal tendencies, escapism, taking refuge (in dreams, daydreaming, alcohol, intoxication, drugs, etc.), sacrifice, unreserved acceptance, caring; ankles, swelling, pain	Neptune and Jupiter

As for the positions of the planets within astrological houses, various recommendations and reports of experiences can be found in literature, among which the position at or near a house cusp is particularly noteworthy. It is supposed to be especially important, giving the year in question a particularly strong emphasis. Moreover, a planet at the very angle (angular positions, i.e. conjunctions with four angular points) is supposed to have a key significance for the whole year, as a kind of motif or key point. As experience and the theoretical model of timing in solar return charts of the next chapter show, the angular element cannot be denied a prominent position, yet it is exaggerated to consider it the main point of a horoscope. A position at the beginning of a house has a function, often as pronounced or as strong as the position in the middle or the end.

Derived houses

The table gives only the basic meanings for interpreting the positions, found in a horoscope, although there are many more meanings. Just by picking up a few astrology textbooks or taking a little more time to browse through the sources on the web, the list of possible interpretations expands enormously. If for no other reason, it is because there is no standard set of concepts, associated with a particular astrological house, and many other connections have been discovered by individual researchers and developers of astrological theory over many years of study.

A very different meaning arises with more complex searches. A sick cat, partner's shares, second spouse or a daughter from her first marriage for whom she is still responsible; there are dozens of such examples to be found at any moment. From the concepts in the above table we can quickly find partner in the seventh house and shares in the eighth. However, this eighth house is our eighth house not his (for the horoscope with the first house indicates us and not him, he is in the seventh house!). That is to say, when looking for our partner's shares, we will start from the seventh house (and consider it to be 'his first'), count up to eight and arrive at our second house. There we will find the partner's shares. This approach is called 'derived houses'.

While the matter can get even much more complicated, let us have a look at a simpler and certainly more common example, i.e. grandparents. If father and mother are signified by the fourth and tenth astrological houses, respectively, this applies to everyone. This means that, as in the example above, we start counting at the fourth house if we are to find the houses corresponding to father's parents, or at the tenth if we are to find his mother's parents. We will arrive at the seventh (father's father) and the first house (father's mother), or at the first (mother's father) and the seventh house (mother's mother).

A small illustration of the "complexity" of the matter is offered by the case of someone fearing for his grandmother's health. If we have "found" the right grandmother (i.e. the father's or mother's mother) following the process described in the previous paragraph, and if we know that health is indicated by the sixth house, we will again count up to six from the house denoting that grandmother. Thus, in the case of the maternal mother, we will find health in the twelfth house of the horoscope, and so on.

These principles are perhaps most useful in the branches of horary and electional astrology, where the precise location of the object or concept being observed is particularly important. But even in the interpretation of a solar return, a basic knowledge of the rules of derived houses is useful. The questions posed to astrologer are very varied, sometimes a bit vague, but the expectations are very clear: "Will Grandma get well?"

However, since the question of parents' health is very common in practice, let us write down a few useful guidelines for studying the problem in the yearly horoscope. If health is found in the sixth house, crises and operations in the eighth, and infirmity or sick leave status in the twelfth, the third, fifth and ninth houses will apply to the mother with regard to this topic, and the ninth, eleventh and third to the father. However, if it happens that our father had left us (during childhood or somewhere along the way), it is often the case that mother "assumes" his place and information about her will be found in the fourth house, generally related to father. This is probably due to the logic, whereby both parents are "home" and are located in the opposite fourth and tenth houses (father and mother), but when one is left alone, he or she is identified with the fourth house, i.e. with home. Thus, in the case of children of single mothers or widows, their health information will also be provided by the ninth, eleventh and third houses.

Repeated positions

We have already mentioned that we use many of the same techniques in interpreting the solar return horoscope as we do in interpreting the natal chart, so interpreting individual aspects of the Sun in the solar return chart seems superfluous. For example, the aspects of the Sun and Mars will indicate a year of being active, combative, incisive, enterprising and all the while at least a little unpleasant, or at least tense and irritable. It is also necessary to examine the rest of the chart, or rather the elements of the chart that are linked to these two planets, because they provide information on the direction of this activity. An opposition between Sun and Mars will bring tension over some unresolved conflict, which can generate anger. A square can bring us fight, a struggle to establish identity or even a battle for survival. Trine and sextile will make it easier to reach a goal. The same goes for the interpretation of this position in the natal chart. Similar rule applies to all other aspects between the Sun and the planets. As for the **transits** of the slow planets, especially Saturn and the trans-saturnian planets, to the natal Sun, the conjunction with such planets also occurs in the solar return. The slower the planet, the more likely it is to be in a very tight conjunction with the Sun in a given year.

Sun in the astrological houses

A survey of the positions of the Sun in astrological houses can be made in several ways, one interesting alternative being astrologer Mary Fortier Shea's approach of listing the positions of the Sun first in cardinal, then in succedent and finally in cadent houses. She advocates this variant by pointing out that the Sun moves clockwise in solar return sequences of three houses per year (i.e. the reverse of the normal direction of movement of the bodies within the horoscope). In fact, due to leap years, the timing of the Sun's position in solar returns changes by about six hours each year, which means that the solar time will fluctuate from year to year. If we do not move from our place of residence over a long period of time and the solar returns are calculated for the same location, we will easily notice the cycle, associated with the movement of the Sun. Thus the Sun will move from, for example, the fourth house of the first solar return in the series to the first house the following year, then to the tenth and to the seventh house, and so on. More specifically, it will move in angular ('cardinal') houses for a few years, then in 'succedent' houses, in 'cadent houses, and then in angular houses again.

Depending on the latitude and degree of the Sun, each of these development cycles can last 10 - 11 years. At our latitude, we can see in practice (for example, by calculating a sufficiently large number of successive solar return horoscopes) that the "three houses a year" rule is only partially true, as the Sun moves two or three houses at a time. To avoid complication, let us present the overview of the Sun's positions by astrological house in the usual way, starting with the first astrological house.

(In reading the descriptions that follow, an astrologer should be aware that these are mainly general descriptions, which may vary from person to person, depending on other elements and highlights in the horoscope, whether or not they are related to the position of the Sun. More on this, together with some examples, will be written at the end of these twelve descriptions.)

Note: When studying the Sun's positions in each house of the annual horoscope, it is important to bear in mind that Mercury and Venus, or at least one of them, are likely to be in the same house as the Sun in that year, and in some years other planets will join in too.

Sun in the 1st house

If we know that the first astrological house primarily means "I", then this will be expressed in some way in the year when the Sun is placed in the first house in the solar return horoscope. It could be a year of personal assertion, progress and achievements. Self-interest is very important in such a year, along with fulfilment of our own needs and expression of our talents. Often there is also a question of a sense of identity, a person seeking recognition and validation, and above all a greater awareness of all that is truly important to them; in such a year we are confronted with the urgency to put ourselves first!

The key motive of the natal Sun is definitely the desire for recognition. This characteristic is largely expressed in the year when the

Sun is placed in the first house in the solar return, especially if it is closest in position to the Ascendant, and particularly highlighted, if there are no planets other than the Sun in that house. One may do something to be proud of in such a year, i.e. realize personal goals and/or gain satisfaction. The Sun's position may, however, be burdened by bad influences from other planets, solar or natal, and then we may experience arrogance, misbehaviour, over-assertiveness, even tyranny, in fact anything from the "badly placed Sun" rubric.

Depending on the rest of the horoscope, especially any other planets in this house, and especially the sign it occupies, including the ascendant sign (if this is not the same as the Sun's), it is possible that the person in question may tend to act independently, perhaps even becoming introvert, should this present a better opportunity for self-expression. As such a year often follows one, in which the Sun is placed in the fourth (home) or third (relationships with others, i.e. something where the 'I' is not so very emphasized) house, the person feels fulfilled and internally content, experiencing many affirmations, but also facing the danger of being blinded by the position.

Sun in the 2nd house

Basically, in a year when the Sun is placed in the second house of solar return horoscope, finances will be a primary concern. It may be about earning money for the first time in life, about a feeling of having to earn more, or about checking one's budget, perhaps to redefine or control costs. We may run out of money (the sign on the cusp of the second house is usually quite informative about our attitude towards money in such a year) and struggle to pay bills. The year brings the urgency to examine our resources, both material and subconscious, which are more difficult to evaluate, for example in terms of our talents. Unless we have developed our potentials sufficiently, life in a year with the Sun in the second house is likely to generate a need to do so. In this year our talents should be flourishing, not be hidden, disbelieved or even limited. This is the time to express and develop them.

Even if our financial problems cannot be solved, we should at least make an effort to put things in order. The situations we will find ourselves in will force us to deal with material problems in a practical way, as well as to become more efficient in handling and dealing with money. It would be best during this period to lay new, firmer foundations for our future path. Last but not least, the second astrological house is related to the principles of the second sign, Taurus, the most typical quality of which is – stability.

More often than monetary themes, however, the other dimensions of the second astrological house are present in the year, i.e. values and, above all, a sense of self-worth. The Sun placement in the second house also indicates validation (of worth or values), a tendency to be valued or to improve one's self-opinion, as well as questioning one's self-worth. Although the mindset in such a year tends to be traditional and conservative, it is usually favourable for a re-evaluation of one's own values and priorities, as well as for a firm decision to get rid of a harmful habit like smoking.

Sun in the 3rd house

This will not be a boring year, but a year full of activity and change. Life will bring us the need to adapt, change, develop fresh ideas, educate, train or graduate. Our curiosity will be greatly increased this year, we will seek and convey information. As befits Geminis, who are analogously related to this astrological house, things can take place in hundred different areas or levels this year, in communication, transport, relationships, information, schooling, writing, analysis, etc. Our mind is focused on the rationality, the approach is intellectual. It is a time for expansion and developing the ability to communicate, as well as to expand our social circle. The term communication relates to short journeys as well as to speaking, writing, reading books, telephone and social networks communication.

The second house sun is also a good time for social activities, especially in the areas pertaining to the third house, i.e. siblings, relatives and neighbours. This is the time to renew or clarify relationships, depending, of course, on other factors in the horoscope.

As the natural ruler of this house, Mercury, is ambivalent, we cannot anticipate only favourable but also unfavourable events in relation to this placement, so a restless mind can also lead to nervousness or depression, but these usually alternate with periods of elation and a feeling of complete mastery over life.

This will certainly be an active year, bringing lots of activity and change.

Sun in the 4th house

In the 4th house, the Sun calls for a redefinition of our psychic, emotional and spiritual relationship to home, security, parents, past and roots. This may point to inner emotional changes, to the relationship with parents (moving away, becoming independent, changing balance in relationship), to dealing with inheritance issues or other (including legal and administrative) activities, relating to real estate, land or possessions. This is a year of a pronounced need for emotional security, and therefore a good time to surround ourselves with people who support our sense of security and nurturing. The year may bring a relocation (which, however, should be supported by some other typical indicators, discussed further in the book), as well as renovation, redecoration or maintenance work. Something to do with home, apartment or office is likely to take up a lot of time, energy and effort this year.

The Sun in the 4th house year often marks a turning point in the perception of one's origins, personal growth and karmic mission. This is a contemplation time, time to listen to the soul. After all reference to the principles of Cancer (the fourth sign of the zodiacal circle) signifies that this is not going to be a year of progress, risk-taking and enterprise. Things from the past are surfacing that need to be addressed and resolved. These may be forgotten or neglected events, relationships, or habits, depending on the respective potentials. This is the year to examine our family priorities, as well as to resolve existential issues regarding our parents and also our children (family planning).

A general characteristic of this solar return year is working on things in retrospective, fixing, cleaning up, tidying up, replacing and clarifying. Possibly some new roots will sprout, bearing fruit in two years' time, when the solar return Sun will probably be in the tenth house (status) or in the eleventh house (focus on future).

Sun in the 5th house

This is the year of being our "true selves", of expressing all our strengths and abilities, leaving others outside. It is the year of making a personal mark, perhaps in a creative or artistic field, time to be proud of yourself and your achievements, and also time to be pleased with yourself. However, as the fifth house signifies many other important areas, the focus of attention and action may be on those we love most, our beloved and our children. A person may give birth to a child (let us not overlook that the potential of parenthood is more visible in the horoscope of a woman and less in that of a man), but in this case it is mainly a matter of identification with that pregnancy or that child; if it is a case of a third child, for example, the fifth house will as a rule not be so highlighted.

In the Sun in the fifth house solar return year, many questions about our lifestyle, personality and areas beyond basic needs and subsistence level are relevant, i.e. do we play a major role in our lives, are we making good use of our potentials, do we know how to enjoy life, are our children more important than ourselves, do we know how to play and have fun and, last but not least, do we love ourselves enough. In this year pleasurable activities are in the forefront. This is the time of sports, competition and rivalry, play, fun, gambling.

Involvement with children awakens the child in us, who we have already forgotten, and we suddenly tend to become more spontaneous and playful again, at least for a little while. Perhaps in some way we are reliving our childhood and youth (and what circumstance is better suited to this than planning or expecting a child?).

It can also be an exciting year in terms of love affairs or changing partners, including sexuality, if this is in any way important to the person in question. In any case, this will not be a calm and peaceful year, but rather one with a lot of drama and personal engagement.

Sun in the 6th house

A highlighted sixth house indicates a prevailing concern with work, service, health or order. Our body may warn us of inconsistencies or mistakes in behaviour (habits and vices are especially typical of the sixth house) or lifestyle. It depends on the personal horoscope and certain additional indications in solar chart, whether we will be facing more serious problems, perhaps even a relapse of a chronic illness, or whether it will be enough to make a drastic change. A highlighted sixth house applies above all to better organization, requiring effort, time and attention. Often in such a year "either-or" dilemmas arise, such as whether our habits are still beneficial to our lives, or whether they may have turned into a duty, an unpleasant routine or even a nuisance over time. Lifestyle mistakes often generate health problems, not only because of poor diet or intemperance - the body tolerates for a while a malfunctioning segment that we do not pay enough attention to, and then warns us with a symptom... If this prompts us to seriously reconsider our lifestyle and habits, we can even thank our body.

In terms of work, this is a year of highlighted important relationships with subordinates or colleagues, of finding another job or trying to correct values or practices in the existing one, and possibly of making major changes. The fundamental question is work itself - are we perhaps overestimating our jobs, tasks and duties? Has the job become more important than ourselves?

A very strong occupation of the sixth house in a yearly horoscope often indicates a general change of lifestyle. Whether these changes are also conditioned by health problems will be discussed further in the book, but this is not a necessity. The sixth house is also strongly highlighted in cases, where there are major, personal changes of residence - for example during or after a relocation, or in the event of someone moving in or out.

Sun in the 7th house

The Sun in the seventh house year is characterized by the importance of others in one's life, either in marital or other types of partnerships, or in making agreements and compromises, in public image, in being pressurized by the outside world or by individuals, as well as not being able to make decisions and take care of oneself. The tendency to cooperate with others, to co-determine, to manage their affairs, to seek their opinion or agreement is very strong. There may or may not be mutual benefit; there may be constraint from others, frustration. In one case the year may bring marriage/partnership, in another case a divorce. Generally speaking, this is the year of learning through others, when someone from our surroundings may serve as a mirror, reflecting our own image.

It is quite possible for a problem to surface, especially after a prior (conscious or unconscious) sweeping of things under carpet, which is quite typical of Libra (the seventh house). Conflict as a result of such a situation makes us aware of our expectations of others, but also of the kind of energy we radiate in our relationships.

This is also the year of going public, of wanting to be accepted, either in terms of prestige or performance or in terms of business connections. Individuals, who are blind to their own faults, are bound to face many problems in such a solar return year.

Sun in the 8th house

As usually, the eighth house is also most difficult to define in solar returns. It definitely denotes a year of extremes, transformation and fundamental change. Something will quit and something new will emerge. We might resign at work and go back to school, radically change our life, come to terms with something and quit for good. Such fundamental upheavals are neither easy nor graceful; pain, suffering, helplessness, fears and tearing apart abound. Only if the Sun and other planets in this house of the yearly horoscope are very well placed, can we expect these changes to be carried out in a satisfactory way and without the principle of total overthrow, helplessness, destruction, or even death.

The eighth house is also the house of instant and deep perception, often psychological or even purely intuitive. We may delve deep, encountering psychology, life research or the occult, thereby altering our perception of other people. The Sun's position, together with the conjunctions, will reveal the extent to which this will be liberating or oppressive, disruptive. For example, we may become aware of manipulation on the part of others, especially a (spouse) partner.

The eighth house is also a financial house, concerning investments, handling money or debt-creditor relationships. For example, such a position of the Sun could signify taking out a new loan or money that someone fails to repay; however, the matter must be sufficiently pronounced, 'worthy' or even pressing, to be indicated by the Sun as the key factor in the yearly horoscope. Often Sun in the eighth house placement relates to a legacy or inheritance.

In general, this position is not favourable, and the year with the Sun in the eighth house will be remembered for anxiety, pressure or fear. Even if the end result is relatively favourable, the path to it is neither smooth nor flower-strewn...

Sun in the 9th house

The Sun in the ninth house of the solar return horoscope relates to expansion, liberation (including independence, although this typically falls under the domain of the eleventh house), knowledge, insight, learning, bringing a year of more positive experiences or greater achievements than the year before. This is a year of searching for new philosophical, religious and liberating directions, time of re-evaluation of our state and way of life.

This is a year of personal growth, bringing more positive energy, enthusiasm, self-confidence and a better understanding of one's own identity. If we believe strongly in ourselves and our worth, it will also be a year of proving our skills and competencies. Otherwise, in a year with the Sun in the ninth house, one can expect to reach a new level of self-awareness in terms of his limits being in fact further than previously believed; sometimes things even happen in a "now or never" kind of way. This is, among other things, also the 'year of search for meaning'. Very often also the year of attainment of a level of education or completion of schooling; the diploma indicator is usually placed in this (higher knowledge) or in the tenth (status) house.

Our curiosity and thirst for knowledge are so strong in such a year that we are very open to all sorts of new insights; we can expect journeys, possibly further afield, changes (usually progress) in our education or intellectual expansion. The ninth house is also related to legal matters, so our focus of attention in such a year may well be some important legal or administrative matter, perhaps obtaining documentation (if sufficiently important enough to be marked in the horoscope in this way).

Sun in the 10th house

This year is very important for the identification of the person in question and his/her public image. He or she becomes aware of his or her role in society, so public image and performance are of paramount importance. The Sun in this house is therefore perfectly placed to achieve career success or to attain higher status, promotion (only to be perceived as fulfilment) and fame. If other indicators are present, this year may bring a change of job, promotion, public praise or reward, even business independence or self-employment.

This is a very good year to make life-changing decisions, depending on the rest of the horoscope; these are not about habits or lifestyle, but about a new direction, usually upwards. Matters with authority figures (boss, job) or parents are also likely. The main focus is on goals to be achieved or defined, perhaps redefined, on realization of our purpose in life or on finding the vocation we were 'born' for. Our aspirations for progress and success become clearer in the year with the 10th house Sun, making us firmer and more determined than before.

If the Sun's placement is not unfavourable (first natal or just solar), such a horoscope can be expected to bring the person a higher or improved status.

Sun in the 11th house

The placement of the Sun in this house is particularly interesting and very different from the others, similar to everything else relating not only to the 11th house, but also to its natural ruler, Uranus, and the analogous sign of Aquarius. In this year everything is questioned, including social norms, rules, established practices and everyday routine. The question "Why not?!" may be the question of the year, especially when it comes to contributing to society, organization, general well-being, and to inspiring others to engage in common ventures and activities.

It is an important year to learn the lesson about maintaining your identity and value even in the case of a larger group, society or organization. The 11th house is also the house of friends and companionship, as well as of alliances and acquaintances that make it easier to achieve our purposes and goals. We may find ourselves facing a challenge regarding friends, wondering what to do with a friendship, we may also make - or break - an important friendship.

The Sun in this house provides us with enough energy to overcome or remove any obstacle (even just intellectual, mental or imaginary). This is a time for redefining goals, plans, perspectives, hopes and heart's desires, a time for vision, perspective, planning ahead and the future in general. Independence and liberation (the rest of the horoscope usually shows us clearly from what) are important highlights now...

The 11th house is so special because it is the only house dealing exclusively with 'tomorrow', something that is yet to come. The Sun in the previous solar return was in the second house, where we have been consolidating or trying to hold on to values, but now it is time to move on, even if we do not know what the future has in store for us. (Also, the probable prior placement in the first house did not concern 'tomorrow', so the year with the Sun in the eleventh house is really the time for a change.)

Sun in the 12th house

For many with the Sun in the 12th house of the solar return chart this placement brings loneliness, solitude or even a tendency to hide, and above all to operate from behind the scenes. Depending on the prevailing active or passive moment primarily in the natal, but also in the solar return horoscope, this may be a year of 'recharging batteries', of withdrawing into a sphere of action, where results are not so visible, but no less important. The focus may also be on exploring (even through oneself) on the one hand (for example, therapy, meditation, self-reflection, inner organization) and on various psychic or physical states involving a feeling of being helpless, hurt, unfulfilled, misunderstood and the like, for example illness (sick leave is a typical theme of such a horoscope), psychological disorders, depression, help-seeking. In such a year it is difficult to define goals, because the world seems full of endless possibilities, resulting in confusion and identity problems. It is common to feel that we could be achieving much more than we actually are, but we neither know the goal nor how to get there. The arising dissatisfaction or sense of loss can generate some form of escape from reality, such as daydreaming, retreating into the unreal world of computers, television or video, or a more concrete escape, such as alcohol. The twelfth house is not a pleasant place, and the rest of the horoscope should be quite favourable to prevent a "bad year".

However, the 12th house is not only about negative things, so active and/or positive placements and conjunctions (e.g. active placements of the Sun or planets and their good aspects) bring personal growth (e.g. taking courses) or the possibility of being successful in our endeavours, although hidden from the public eye. Such a year is excellent for lobbying, research and any other activity, carried out in a closed or remote setting, a good example being the start of therapeutic work.

Usually the Sun is in the ninth house in the next solar return, bringing information on how to get out of this state on the one hand, but more likely also a sense of release, as well as end of containment and limitations (the fundamental meaning of the ninth house being expansion). However, the Sun may also be in the tenth house in the next year, showing that the client has used the year perfectly to recharge his batteries, which in turn will prove useful in the next year of higher aspirations.

A few more practical observations

Descriptions of the Sun's position in astrological houses are, of course, quite general, but in practice very different cases can be encountered. Thus, one may come across the case of a solitary placement of the Sun in the eighth house; although it does not bring about a 'dress rehearsal', the person may take up some other activity, indicated in the yearly horoscope, with a Plutonian approach, i.e. radically, efficiently and "brutally" - once and for all. The Sun's placement in the eighth house gives the person a new *modus operandi*, one he/she is not used to, but which is necessary for some specific reason in this year.

Regarding the "advantages" and "disadvantages" of the Sun's placement, the principle of the 'natal premise' applies in full, meaning simply that even a favourable planetary position in a solar return cannot generate a highly favourable effect if that very planet is placed very unfavourably in the natal chart, and vice versa. If we take the Sun's placement in the sixth house as an example, and it turns out that the person has a problem with a job or wants a new one, a "favourable" Sun quality brings an active attitude and relatively good potential to resolve the unpleasant situation; an "unfavourable" one, however, shows that although the area is highly stressed this year, the person is not in control of it, spinning around in search of right information or outcome, not necessarily finding one. If health is the matter of concern with the same placement (sixth house), an "unfavourable" position suggests illness and more or less serious problems, while a "favourable" one points to problems that can be solved without too much trouble and strain, or even to dealing (possibly successfully) with another person's health.

In practice, there are also cases where natal chart is predominantly - i.e. not radically - unfavourable regarding the Sun (the same applies to other planets and points in the horoscope), while the solar horoscope is strongly favourable. This is where the assessment of possibilities, that astrologers are constantly faced with, comes into play.

Calculating a long enough series of solar returns, we can see that the Sun appears in the same house four years later, and then (most likely) twice four years later, which means four times in a row at the same intervals. Two years later it can be found in the tenth house, and the same four years after that, while in the next "round", and certainly in the fourth, it already moves into the eleventh house. So at the end of this series Sun shifts into the house of perspective and forward looking, after having been focused on status and building up for a whole series of years. The Sun then does not return to the fourth house for a number of years, but when it does, it will again do so four times.

To avoid misunderstandings or misjudgements it is necessary to point out the involvement of the two sets of planets, i.e. natal and solar. For example, if the Sun forms two squares and one trine with other planets in the natal chart, these will also be found on the outside of the solar return horoscope. This means that some of the aspects, formed by the Sun in the examined year, will be practically "permanent", since they appear in every yearly horoscope. For this reason we cannot claim that these two squares and the trine contribute significantly to the Sun's aspectedness and to the assessment of the "auspiciousness" of the year. They do not convey any new information, so the assessment of the Sun's impact, as compared to the natal premise, will have to be based on examination of the aspects the Sun forms within the solar return chart itself, i.e. with the 'solar' planets. More detailed information will be provided in the further chapters, including the particular combinations of highlighted houses in the solar return horoscope, which will help us reveal the areas in focus, for example work versus health, both of which are associated with the sixth astrological house.

PLANETS IN SOLAR RETURN CHART

Unlike the Sun's position, which is the starting point for calculating a solar return horoscope, and is by the nature of things always exactly the same within the birth zodiac sign, the other elements of the horoscope are different. They are found at different points, and very different patterns of movement can be observed. The slow planets, from Jupiter onwards, move around the zodiacal circle in a similar way as if they were observed in transit charts. Saturn will thus stay in the same sign for three consecutive solar returns, but each year closer to the end of the sign. Uranus will keep in the same sign on average of seven times, Neptune and Pluto several more. In the case of the latter we must not forget the elliptic form of its orbit, which is why it occupies a sign for a period of between 12 and 24 years. Because of its "slowness" we can expect it to stick around the same degree for two or three years, depending on the phase of directness or retrogradation, still within the orb that is taken into account when interpreting the solar return horoscope. To put it simply: if the position of solar Pluto is in exact conjunction with the Sun, then it was also conjunct the Sun in one solar return before, and will be in the next one. Saturn, on the other hand, will be in conjunction with Sun in one or at most two consecutive solar returns.

An extensive research paper could be written on the movement of the individual planets through the sequences of solar return horoscopes, but for the present book, let us only be precise enough to say that in each successive solar return horoscope we will find Jupiter almost exactly one sign further ahead in the zodiac, given that its orbital time around the Sun is 11.86 Earth years. Saturn will move 10 to 13 degrees in each year. Uranus also moves in similar steps, except that in the 60s (when it was in Leo and Virgo) this was about 5 degrees, in the 80s (Capricorn) a good 4 degrees, and more recently (Pisces, Aries and Taurus) about 3.5 degrees. Neptune's motion is much steadier, moving predictably from horoscope to horoscope by about 2°10' forward all the time. Pluto's movements in the 60s (through Virgo) and today (through Capricorn and Aquarius) are practically the same as Neptune's from year to year, but after Scorpio (in the 80s) it has been moving for more than 2.5 degrees from year to year. (His stay in his own sign is among the shortest, though.)

The personal planets are different in this respect. Mercury and Venus are always somewhere close to the Sun, so when observing solar return sequences, they will never be found more than one (Mercury) or at most two (Venus) signs ahead or back from the Sun's position. The same applies to the house positions where they move close to the Sun and are no more than 1 or 2 houses away from its position, respectively. However, even their positions over a certain number of years can be found in all the 12 houses of the solar return horoscopes, which is not the case with the signs - Mercury is only found in a maximum of three signs (the Sun's and two adjacent ones), and Venus ideally (if the Sun is positioned at exactly 15° of a sign) in five signs.

CIRCULAR RIDE AROUND THE ECLIPTIC

The influence of the signs is examined in many astrology books and manuals, but in our case it is important to note that these are linked to the associated astrological houses in the yearly horoscope, varying with every year. This means that the different areas of our lives, i.e. partnerships, finances or schooling, also undergo changes in successive years, all in accordance with the signs.

The sign of **Aries,** which will be in a different house each time in the yearly horoscopes, will bring tension, struggle, striving for independence, recognition, identification in the pertained area. *(Quite a bit of variation in all cases may be brought about by a Malefic, placed in that or a particular sign, perhaps even in some stressful aspects with other planets, but we will not elaborate on that here, as the basic* *information would be lost.)* The sign of **Taurus** will bring routine, habits, a less strenuous area, perhaps connections with aesthetics, nature or finance, but generally not in a negative connotation. We will seek or achieve stability in this area. The sign of **the Gemini** will be related to change and communication of any kind, possibly to dispersal or even complications, but certainly a lot will be going on. There will be no peace in the respective area, but progress or development are possible. The sign of **Cancer** will bring stagnation, backtracking, spinning in place, procrastination, but also a struggle for existence. Its respective sphere will provide neither development nor progress, these will have to be sought in one of the coming solar returns. The presence of **Leo** will add some drama and tension to the respective field of the yearly horoscope, and while it will be of a creative, competitive or managerial kind, it will not automatically be favourable, pleasant or helpful. Moreover, in this area we may be inclined to seek uniqueness, glitter and praise. With **Virgo** the first half of our ecliptic journey has been completed, which is long enough for anyone to notice the alternation of active and passive or 'favourable' and 'unfavourable' years. Virgo will bring boredom, routine, submission, stagnation, an attempt to keep this area under control, and retreat rather than progress to a particular house of solar return. We will tend to establish order in the respective sphere, but will not experience any drama or misfortune.

The second half of the zodiac circle brings six principles and meanings that are opposite to those of the first six signs. **Libra** is also a cardinal, major sign, belonging to the active and enterprising signs, yet its activity is expressed in a very different way. Since it favours agreement, consensus, partnership and harmony, it will also bring a kind of balance, non-aggression, well-being in the solar return. **Scorpio,** on the contrary, will tend towards radicalism, which many of us dislike, so the associated area in solar return horoscope will be defined by a crisis, a problem, a battle, fear or powerlessness, perhaps a break up or end. Nothing pleasant for sure, unless we want to destroy something and have so far failed to do so; in that case a year with such a horoscope will be welcome… The area indicated by the presence of **Sagittarius** in our yearly horoscope will be characterized by expansion, breathing, freedom (of expression or action), perhaps we will learn a lesson in this area, or things will "finally" start going. **Capricorn** will bring rigidity, sobriety, complexes, constructiveness, business and/or goal-orientation, but definitely a standstill or hindering rather than progress. **Aquarius** demands liberation, independence, the house in the yearly horoscope associated with it represents the area, where something will be different from the past, or entirely new, unknown, but definitely moving towards self-reliance, independent action. **Pisces** will refer to inspiration, obscurity, the sense of being lost. The compass will go in several directions or nowhere in particular - we will have everything and nothing in this area, wondering what is actually going on and waiting for better times.

THE MOON

The Moon is always one of the most important elements in an astrological chart, be it in a female or a male horoscope, in a strong or weaker position, playing a strong or a ruling role in the chart. Symbolically, the Moon rules water, which makes up 70 per cent of our bodies, and if it moves immeasurable quantities of water around the world daily in the form of tides, it certainly impacts humans too. In a solar return chart, the Moon is in a different place every time, showing the emotional part of our personality, more broadly, our *yin* half. Certainly the point, where the Moon is positioned within the yearly horoscope, is considered most crucial, whether we are looking at its position through sign, house or aspects with other elements of the horoscope.

The house in the solar return horoscope containing the Moon will point to the area of sensitivity and change in the year in question. Amongst various other modes of action (motherhood, water, emotions, etc.), the Moon also resonates with the principle of flowing water, flux, fluidity. We will also be sensitive and vulnerable in this area, our emotional needs being noticed and highlighted. The house placement of the Moon is also associated with dependence on someone or something. Related personal qualities may be sensitivity, receptivity and protectiveness (of others), as well as restlessness and defensiveness.

This principle of flow and sensitivity of the Moon is perhaps best illustrated by the example of the two financial houses. The Moon in the second house usually means the flow of money, one could say *"a lot of money will pass through (this person's) hands, but it will not necessarily stay with the person"*. The degree of emotional perception here depends on the Moon's sign, i.e. the sign in which it is placed, the sign on the cusp of the second house (if not the same sign), and on the Moon's aspects. It is quite possible that all the money flowing in will not particularly touch the person emotionally, being related to planned, expected or less important sources. In any case, the client can be told that "there will definitely be money", because an outflow automatically means that there will also be an inflow, otherwise the financial situation in this year would be marked by the presence of, for example, Saturn - there is less and less money, or none at all...

The opposite eighth house is a little more complicated to interpret. Although financial themes are also possible (investments, income from so-called 'foreign sources' - insurance premiums, rents, lending and borrowing money, etc.), along with or inheritance issues, there are in fact more possibilities and, above all, the emotional involvement is much more intense. For example, there may be partner's money or partner's transactions with joint money, which rarely allows for emotional distance. This is especially the case when it comes to shifts in family values and relationships (inheritance and legacies), and even more so when it comes to the issue of the mother or the wife (one of the Moon's essential meanings, especially in a masculine horoscope).

The Moon's position is much more likely to be perceived emotionally, with unrest or even fear, in the eighth house than in the second house – unless, of course, it relates to those born in Taurus or those with strong financial attachment; an astrologer needs to know his client well enough to be able to forecast such possibilities.

Relationships with mother or wife also require special mention when analyzing the Moon in solar return; if we find that the Moon no longer symbolizes mother in a man's horoscope, or if mother is deceased, the Moon relates to wife. Depending on the sign of the Moon (the type or mode of relationship) and the astrological house, we can infer to what extent and in what ways the person's mother (or wife) will be important in the solar return year, as well as how much emotional involvement there will be in the relationship (especially if there has not been any intense activity in the relationship in the years before), be it about happiness, support, elation, hurt, discomfort, or fear. The relationship does not have to be interactive, i.e. involving concrete events, it can just be just about a feeling, without any real activity.

Moon in the signs

If the Moon is placed in **Aries** in a given year, we can anticipate rapid responding and spontaneous, open reactions, without much self-control or calculating. There will not be much diplomacy in such a year, circumstances leading in the direction of making decisions without hesitation. This is also the time for taking the initiative in emotional relationships and making new contacts with people. Intuition should be at its height in such a year.

A year with the Moon in **Taurus** brings increased need of emotional stability and predictability of events. Not infrequently, there may also be an increased need to improve one's material situation and comfort. This can also turn out to be disadvantageous, small dietary indulgences generating increased weight, especially if the Moon is positioned in the first, second or sixth house, or if it is also the leading planet. In a way, such a year can also be about "getting back to nature".

The Moon in **Gemini** is all about change, activity, state altering. Perhaps not exactly shocking changes, but not necessarily welcome ones. The year brings intense curiosity, yearning for new experiences, places and people, new relationships, communication, analysis. In general, the year is characterized by emotional distance and a distinctly mental approach to life.

The Moon in **Cancer** is such a powerful influence that its house placement in the solar return horoscope is of lesser importance. In any case, it brings a very emotional year, a desire to be protected, but also to protect others; a tendency to belong, sometimes even to the point of dependence. The urge to be part of a whole, a society, a cell, an organization, providing a sense of security, will be highly pronounced. Also typical are search for safety, sensitivity and hypersensitivity. This is not a year of achievement and rapid progress, but rather a year of small satisfactions, sometimes on the verge of minimalism.

In a year with the Moon in **Leo** the need to express oneself and one's worth will be very pronounced. There will be great need of attention, recognition and praise. The desire to be the centre of attention will be pronounced even in normally modest people. Generally, this is a creative (not necessarily artistic) and dramatic year, with a lot of noise and activity, with good results being possible, but not easily achieved.

The Moon in **Virgo** brings a more reserved, analytical, thoughtful, calculating approach and a highly organized everyday life. Issues of health, well-being, food and fitness will be top priorities.

The Moon in **Libra** brings the need of harmony and understanding, when one is not focused solely on one's ego, but capable of seeing both sides of the coin. The willingness to agree and to be agreed with will be strong. Of course, it can also be a year of romance, of enjoying life as a couple, of aesthetic pursuits, of interior decoration, etc.

The Moon in **Scorpio** signifies depth and intensity of experience. We will be serious and determined about our feelings, cutting through and eliminating dilemmas that may arise. A very radical approach will be evident. If other indicators (e.g. other predictive techniques) are favourable, there may even be significant shifts in personal growth, usually related to emotions or hidden intentions, expectations, fears. Complicated relationships with mother (possibly also with wife or some other significant female figure) are very likely. A resolution of such a relationship is welcome, however painful, and it may well be just one stage in the resolution of an – often karmic – relationship.

The Moon in **Sagittarius** generally signifies a desire for freedom, ease, unburdening, knowledge, travel, foreign worlds. We may expand our horizons, which may be manifested not only in study but also in philosophy or religion. We may be too quick to judge, but we usually manage without too much trouble. If the Moon is placed in the first, second or sixth house, increased weight may be an issue in this year.

A year with the Moon in **Capricorn** in the solar return horoscope brings emotional restraint, problems with or because of mother, calculation, caution and sobriety. Goals will be clearly set, but at the same time there will be an awareness that they will not be easy to reach and will require a certain amount of renunciation or sacrifice to be realized. There may be a feeling of not having achieved everything, which may again lead to new goals to be fulfilled in this or in the subsequent year. There is also a need to balance business and personal life, such as career and motherhood.

Independence and autonomy will be very pronounced in a year marked by the Moon in **Aquarius**. We want to do things differently or our own way, routine and orders irritate us, and we are quick to take a 'trip off the beaten track' or to experiment. If independence from maternal or other restraining influence is in the offing, the placement comes in handy. Idealistic thinking can generate a changed rhythm of life. Conviviality and group activities will also be highlighted.

The Moon in **Pisces** marks a 'year of sensitivity', bringing understanding, compassion and lack of determination. We will be looking for stable ground under our feet, which keeps slipping away. If we interpret 'higher' levels, the year may be about seeking the mystical or spiritual, however, depression or seeking refuge in tears, alcohol or something even more concrete, are also an option.

Moon in the astrological houses

The Moon in the **1st** house brings increased sensitivity, receptivity, susceptibility to influence, irritability, instability and mood swings, especially if it is also the leading planet of the solar return. We are highly reactive to the surrounding atmosphere, as well as sensitive to injustice and emotional suffering of people, especially if they are close to us. The importance and influence of the mother is more pronounced or heightened in such a year, and there may also be a transference of experience or a kind of identification with her, or perhaps a preference for her as opposed to the father.

In a year with the Moon in the **2nd** house, there will be a marked increase in the need for security, not so much through retreat or sheltering, but through a desire to become financially secure. This position of the Moon intertwines our emotional and material perceptions, with a full wallet as an effective guarantee. As already mentioned, this Moon brings a 'flow of money' - its inflow does not dry up, but it also runs in the counter direction. If we had serious financial problems in the year before, now is the time to 'fill in the holes'. Also, the instinct to make money will be stronger this year, but we will also be prone to disappointment should the earnings fail to be proportionate with the invested work.

With the Moon in the **3rd** house, the main focus is on activity and changes in relationships, both with siblings and relatives, as well as with the immediate environment, such as neighbours. Combined with certain other elements and highlights, such a year can also bring the potential for relocation (discussed in more detail further in the book). Our curiosity and communicativeness are also heightened, this being the year of flexible, adaptable and active mind.

The Moon in the **4th** house is basically similar to the Moon in Cancer - a position so strong that it can be considered one of the strongest in a yearly horoscope. This is certainly a year when emotional security, home, parents, domestic matters, such as real estate deals or renovating the house, are highlighted. We will be focused on the past, so this is not the year of advancement, breakthroughs and successes, but rather a year of contemplation and, above all, clarification, including some counselling or therapy, especially if reasons can be traced back to childhood or youth. As with the previous position (third house), this placement, in the right combination, can signify relocation. Our preferred location this year, though, will be home.

The Moon in the **5th** house refers to creative potential, sometimes also to education (but even more so teaching), and/or sport. But it is especially indicative of matters of the heart, changes in love or partnership. Given the Moon's nature, we can expect more activity or even more people involved, rather than just one beginning or ending love story; even if there is only one relationship, it may undergo several changes. Emotions will be expressed more loudly and dramatically this year. The Moon in the fifth house is also intrinsically linked to motherhood and all issues relating to children, especially in cases of first-time motherhood.

The Moon's impact in the **6th** house can be reflected either in the area of health and/or lifestyle habits, or in the area of work, service or relationships with subordinates or colleagues. When relocation is indicated in the horoscope, this Moon placement is an important prediction confirmation, every move to a new place bringing lifestyle changes. In the case of a job, such a Moon usually does not bring a change, unless it is in company or conjunction with some element that more definitely points in that direction. A more probable effect of a sixth house Moon is adjustment to either colleagues or to different working conditions, emotional perception being an inevitable companion to this placement. Otherwise, the Moon in the sixth house brings a concern for order and "putting things in their place", analysis, a methodical approach and organizing activities.

In the **7th** house Moon year, partnership will be a major focus. Even without a partnership relation we will be focused on the ones we share our time and efforts with, such as business partners or the public if these are important in our lives. Again, there will be many changes in quality of relationships, agreements, compromises... In the case of marital problems, this situation, if not burdened by some very unfavourable aspects, usually means an attempt (or attempts) to reconcile or to mend the relationship, not a break-up or an end to marriage.

The Moon in the **8th** house year is referred to by many authors as 'the year when intimacy and sexuality are highly emphasised'. This is not to be doubted, however, it depends on the perceptions and priorities of individual, for the Moon in an uncomfortable and often complicated eighth house can have many other meanings. Some may be of financial nature (foreign or joint money, investments, bills, credit, etc.), however, a higher level can also mean 'finding out the true value of others', as in the case of being let down by a close person (otherwise no emotions would be involved - the Moon) and having to adjust internally. What is more, the Moon - especially if there are more planets in this house - indicates the potential for some kind of inner dress rehearsal or deep insights that have been accumulating over many years. A Moon so placed can also indicate health-related events (in combination with some other indicators), for example, a surgical procedure, an injury, and so on.

The **9th** house Moon can indicate a journey (or several journeys), especially if travel has been our heart's desire for a long time. Another frequent possibility, related to this position, is a change in study, for example, resuming after a latent period or considering a change of the field of study. If a thesis is in preparation or progress, then the time marked by such a Moon is usually very fruitful and action packed. The Moon in the ninth house is also often involved in divorce and court proceedings, where sometimes matters stagnate or drag on, and then suddenly activity is resumed. The Moon in itself does not bring a bad outcome here, however, the latter is possible in conjunction with other indicative elements.

The **10th** house Moon indicates a period when our perception of our social or business status will be highlighted. Acceptance and respect will be important, and we will strive to improve our status. This Moon also brings awareness of the needs of the public and those who govern and administer. In terms of work (and career) it brings change within the existing system rather than a change of job.

The Moon in the **11th** house will bring developments in the area of friendship and society, perhaps some organized pursuits, but also an aspiration for diversity, independence or a new path. This placement often brings new perspectives or vision, alliances and friendships, even chance encounters, helping us chart the desired direction. Someone striving for independence (from a relationship, home environment or habit/practice), will find this Moon placement more than welcome.

With the **12th** house not being a favourable placement for planets as a rule, this Moon position in annual horoscope will not bring much good either. Isolation, taking refuge (retreating to a safe environment, even if it be a dark room or alcohol), depression, as well as feeling isolated and helpless are very typical of this lunar position. If favourably placed (conjunction with other elements), this Moon can mean self-care, yoga, meditation, contemplation, even work behind the curtains, if not, it is advisable to be able to cry, get lots of sleep and recharge our batteries for the times when strength will be needed. In terms of health, this position indicates medical check-ups or sick leaves, less often crisis, serious illness or surgery.

More about the Moon

A strongly placed Moon usually indicates a desire for change. Even if involved in "lighter" aspects with other elements, it testifies of restlessness and a tendency towards diverse experiences. It is true that in such case it sometimes does not go beyond the level of a change of hairstyle, the purchase of new clothing or a decision to lose weight, however, the person will nevertheless be dissatisfied if everything remains the same. If there is a more prominent position of the Moon, the changes will be greater or more long-term, the placement may open up a trend, and perhaps the person will simply become more popular this year. The more aspected the Moon, the stronger the potential indicated, as more of the horoscope is involved. Even with unfavourable aspects of the Moon, the goals and desires can be fulfilled, although with more difficulties or stress, perhaps due to over-ambition or not being confident about one's goals and potentials.

Moon in aspects

With positive aspects between the Moon and the Sun, we may experience some advantage, pleasure, joy, fulfilment. A positive aspect highlights the inner balance between *yin* and *yang*. Often the focus is on the man – woman relationship, usually a satisfying one. If a positive aspect is found in a birth chart, we can infer a positive and constructive relationship between the parents, i.e. a favourable environment for the child to grow and develop in. The parents had a proper division of roles and a balanced upbringing, providing a secure environment for the child. In a yearly horoscope such an aspect shows cooperation, agreement, proper division of roles, harmony and success. Depending on the prior or several prior annual horoscopes, such an aspect can also indicate "only" an improvement in an otherwise not fully satisfying relationship. If astrological charts are indicative of our perception, the question is why such an aspect would be prominent in a permanently satisfying relationship; it would be no different in the current year, there being no need for it, since it would not bring anything new. Therefore, if an aspect is noted in an astrological chart, it certainly has a meaning that is worth exploring.

The opposite is the case with a stressful aspect between the Moon and the Sun. As in the previous case, these are very often "male-female" themes, but in a very different light. Here we will find more or less disturbing inner tensions or relationship conflicts.

Mercury aspects suggest a link between mind and communication, as well as the faculties of memory, concentration and objectivity. In the case of bad aspects, problems can arise from being overly critical and sensitive to the words spoken.

Venus aspects always show a loving nature and affiliation. A good aspect is indicative of harmony at home and of happiness, also of predictability and consequently greater (emotional) security. Troublesome aspects indicate emotional problems or wound healing.

Contacts with Mars highlight individuality, which can be aggressive in the case of a conjunction, and inflexible or overly sensitive in the case of an inconjunction. In dealings with others, as well as with projects and ideas, this can be quite a passionate time, but if they are to be carried through to completion, some discipline will certainly be needed. Stressful aspects can also generate domestic conflicts or even sudden health problems.

Aspects with Jupiter are generally favourable, bringing contentment, new opportunities and clear horizons, as well as generosity and kindness. They can inspire confidence in others. A bad aspect rarely brings anything very unpleasant (except if the planets are also negatively aspected in natal horoscope), however, there may be impracticality, overlooking something important, underestimating a challenge, or perhaps making a wrong or subjective (emotionally based) decision.

The Moon's contact with Saturn brings a focus on serious issues at the time denoted in annual horoscope. A sense of responsibility, loyalty and fidelity are highlighted, with no time or space for play. A conjunction brings frustration, a feeling of being tied down, or some uncomfortable phase in a relationship with a partner (in a male horoscope) or mother. A diminished sense of self-esteem is not surprising.

The Moon's aspect to Uranus in a solar return chart 'brings sudden emotional reactions, events or upheavals, something to do with change or independence, suddenly and with a strong emotional charge. Perhaps it could be called a strong tendency to break the shackles. With a positive aspect, however, excellent and original ideas may emerge.

Contacts with Neptune bring sensitivity to stimuli, often idealism. Bad aspects generate vagueness, fogginess and a feeling of being lost or caught in a vicious circle.

Aspects with Pluto are rarely favourable, but if they are, they allow for a breakthrough in an activity or an idea that has long been waiting to be realized. However, a volcano or a geyser syndrome is more likely, especially if the person has been accumulating tension for a long time and nothing could be done. Conjunction also brings the ability to pressurize others, along with severe intolerance. With aspects to Pluto, perhaps even more than with other planets, it is well worth to check whether the same aspect is also present in the natal horoscope; in this case the Pluto solar aspect may also be one of the main influences in the respective year.

LUNAR NODES IN SOLAR RETURN

In addition to the two sets of planets, the combined solar return chart also contains two pairs of lunar nodes, natal and solar. These are not two bodies, but mathematically calculated points on the ecliptic, which lie exactly opposite each other in horoscope (i.e. in opposite signs), and which astrology closely links to the principles of karma and the chain of lives. To recapitulate in the briefest possible way:

Lunar nodes help us answer the following questions: *why am I here and what is my mission? What should I be doing with my life? What am I doing on this globe in the middle of the universe at this time?* It could be said that the North and South nodes encode our life purpose.

The Northern Node is our destiny and karmic path. It resembles a climb up a mountain that seems to be a necessity and a success at the same time, but the climb is not in the least easy. It is the path of learning.

The South Node shows our innate gifts and past lives. It reveals the gifts and talents we had brought to this life, our "advantage", our comfort zone. In these areas of life we are probably already inherently good and successful, even without too much effort, and we may even experience the feeling of *"been there, done that"*.

The demands, implied in the North Node placement, require us to move out of our comfort zone. If and when we do, we will be surprised by the feeling of fulfilment. This is the activation of a life mission.

However, people often and easily gravitate towards the potentials of their South Node, as it resembles coming home to our roots, to an easier path. We may even believe we "should" stick to the South node just because we are good at its principles, but that is not our karmic task.

The South Node should serve as a kind of "a springboard" in fulfilling one's karmic mission.

The meaning of Lunar Nodes

The annual horoscope is, of course, only a provisional horoscope, valid only one year of our life. In this context, the lunar nodes cannot be interpreted as "fulfilling of our karma" - every year anew. The interpretation of these positions must therefore be defined in two ways:
- Position in the annual chart *per se*
- Link to a natal chart that is part of a combined chart.

The South Node will point to an event or an experience in the year when we will only be able to say: *"This is the way it had to be, nothing could be done here"*. The house placement and aspects with other elements, especially the solar ones, will tell us what this finding was related to. Sometimes we will simply be relieved by a fateful event, in other cases we will feel a sense of betrayal or loss. The more closely the node is aspected to the planets, the more intense our experience of the event will be; it will probably also be possible to relate the experience to the other events of the year in question.

The North Node points to a challenge, to something we should do, achieve or overcome, or to something that presented a problem, an obstacle. There is no guarantee of success in this, but the position of the North Node and its aspects will provide more information on the nature of the challenge, as well as the likelihood of victory versus failure.

Some examples: North Node on the Midheaven (MC) may bring the challenge of job change or gathering courage to make a fresh start, even though this may prove difficult. South Node on the Ascendant, though, may bring the realization that a character trait we are well aware of and identify with (Ascendant = self), but are either ashamed of or have tried many times to get rid of, is still too difficult to deal with. The North Node in the fourth house often indicates a decision to move away and become independent, even if we do so with sadness or fear of betraying someone. The South Node in the eighth house can indicate the awareness that a close person has badly disappointed us; we may now see this person in a different light, yet nothing can be done. (In this manner, dozens of practical interpretations of the Moon nodes are to be found.)

However, when the Nodes are strongly related to the natal horoscope, the events they indicate together or separately, will be one important station in the development of life's mission, unfolding right in this very year; in other years with less intense aspects, though, the Nodes can be interpreted as described in the preceding paragraphs. Especially aspects with the natal Nodes, the Ascendant, the luminaries, the ruler of the Ascendant, Saturn, etc., point to a karmic task, awaiting us in the present lifetime.

Here the North Node interplay may show a part of fulfilment, perhaps an important step in life, while the South Node may be about realizing that the potentials it brings are not the right path, and that something will have to go if we are to fulfil the life task, encoded in the horoscope.

The Lunar Nodes are another special feature of solar horoscopes, concerning primarily recognition and perception of an event, rather than the event itself. Very often, at the point in time indicated by the Node, there is only the awareness of something having to be done or changed, with the conditions not being ripe for the respective change just then. It will have to be carried out when possible, although the decision to make that change at the point, indicated by a Node is clear, firm and final. This phenomenon is particularly frequent in job changing, but also in partnership that no longer suit us - *"I can't do anything right now, but I'll do it at the first opportunity!"*

Often in yearly horoscopes the Lunar Nodes are closely linked, or even placed at the point in time when we lost someone, not as a sudden loss, but as the end of a process we have been expecting, even if we may not have admitted it to ourselves. The presence of aspects with Saturn or Pluto, especially inconjunctions, are a fairly strong indicator of the loss of a close or familiar person.

THE IMPORTANCE AND INFLUENCE OF THE LEADING PLANET

The term "leading planet" has not yet become familiar in our astrology practice, although it is neither new nor unusual. It is most easily described as "the first planet from the Ascendant onwards" in the first house or beyond. As such, it in a sense "leads" the planetary family in the horoscope in question, which can also be illustrated by the concept of "locomotive", familiar from the theory of planetary configurations.

Its significance is logical, originating from astrological tradition, which attributes greater accidental dignity to angular planets, angularity being a kind of "sacred" position. The angular planets are, of course, related to the four astrological houses, where greatest power is attributed to the first and tenth house; the term 'leading planet', however, is much narrower, since it only denotes a planet in conjunction with the first, most important angle, the Ascendant.

From here onwards, astrological practice knows a few deviations. Some authors argue that only a planet in the first house can be called 'a leading planet', regardless of whether it is placed close to the Ascendant or at the very end of the first house, while others consider a planet in the second or even third house to be 'the leading planet' as well, but only if there is no planet between the Ascendant and the respective planet. The leading planet should definitely 'be placed in the first quadrant'.

To understand the kind of information the individual planets in the 'leading' roles might contribute to the interpretation of solar return horoscope, we need to take into account the strength of the planet in question. This is influenced by various factors, such as the dignity of the planet and its aspects to other elements of the horoscope. Thus, for example, Saturn placed on the very Ascendant of the solar return, does not necessarily bring only problems and misfortune, if its role in the natal horoscope is favourable or supportive. There will certainly be some important clarification, setback or lesson, proving to be crucial and beneficial for life later in the year. As with all other planets, interpretation of the leading planet depends primarily on the potentials of the natal chart.

Let us now take a look at the interpretation information each planet provides in the role of the leading planet in a solar return horoscope.

The Sun, which, like Jupiter, mostly brings favourable developments, will be the leading planet in terms of energy, dynamism, perhaps even brilliance, making the year in question a year to be remembered. Depending on its aspects with other planets, it is quite possible that the activities, associated with this position, will not only be favourable, drama being more important than pleasure, kindness or a positive outcome. This is especially true if the Sun of the person in question is placed in a problematic or difficult position in the natal horoscope. Such a Sun can also bring a greater focus on relationships with the masculine principle, especially the father, but also with authority figures in general.

As the 'leading Sun can also be placed in the second or third astrological house, its influence will be manifested there, either in terms of a struggle for recognition, validation or independence (financial or otherwise), or in the dramatization of the financial situation if in the second house, as well as in the area of relationships, communications, change if in the third astrological house. Cases where such an event or development becomes significant not only for the time being, but widely, are not infrequent.

If the first planet following the Ascendant is the **Moon**, many changes, adjustments, a constant flow of activity can be expected in the respective solar return year. One's sensitivity will be very high in this year, the Moon itself signifying a strong theme of motherhood, home, security, family, including pregnancy, which must be very important for the person, either because of its significance (for example, the first pregnancy, i.e. the status of motherhood, or a very late pregnancy after many years), or because of its specificity, such as a risky pregnancy, which in itself brings a very delicate period. With the Moon as the leading planet, we can also anticipate various heart desires or plans for travel, change, lifestyle and so on.

Mercury as the leading planet brings diversity, interaction, communication, lots of activity, agreements, meetings. There may be news of change, which should be neither bad nor restrictive (unless Mercury is in close conjunction with Saturn, for example), even if Mercury is badly placed. In this case, the news will not be immediately useful or will be misleading, giving rise to expectations, illusions or disbelief, but certainly not bringing disaster. There may be desires and plans for travel, regardless of destinations or possibilities.

Mercury in the first house often signifies youthfulness, because as the leading planet it can bring a sensation of youthfulness even at a time, when we have already stopped feeling young, for example after a meeting with schoolmates.

Expectations from **Venus** as the leading planet are usually highly overrated. Venus is traditionally the "Lesser Benefic", bringing gifts, joy, pleasure, beauty, even profit, not to the extent of a definitive breakthrough upwards, but only at the level of satisfaction. Even if placed very favourably in its leadership role, it only brings something relatively pleasant in the current year, perhaps making us feel good about ourselves or experience some major confirmation (this applies to both first and second house placements, however, with some differences in interpretation), but not bringing permanent happiness or a positive resolution of something we have been experiencing in the times before.

If, by contrast, Venus is negatively placed or in an unfavourable natal position, 'its leadership role might even be unfavourable, bringing deception, profit-seeking, blackmail, forgery or similar. However, since Venus also rules love and younger women, these areas may well be highlighted in the Venus leadership year.

A specific and not uncommon explanation of this placement is the feeling of satisfaction with oneself or the success/development of a situation, in case the situation had been unpleasant, depressing or problematic in the time **before** the current birthday. When the period is over, we often end up pleased with ourselves and feeling good.

If the first planet following the Ascendant is **Mars**, the year will be highly dynamic and full of conflicts, clashes, confrontations - unless Mars is without any bad aspects. In this case, it can be attributed combativeness, struggle for self-assertion and victory, these being the key focuses of the current year, along with the entrepreneurial spirit and the launching of new projects.

As Mars rules two signs and (in each, but also in the annual horoscope) at least two houses, the probability of its ruling one of the health-related houses (6th, 8th and 12th) is quite high, associating its rule in an unpleasant, painful, suffering way with the first house, which represents our physical body; in this case, it can be said that the health segment will be very pronounced in the year under observation.

This conclusion, however, is irrelevant with Mars being placed in the second or third house of solar return as the leading planet. Mars in this role may also imply the presence of the masculine principle in the year, if it was absent or very weakly expressed in the prior years; however, this applies only to female horoscopes and to the charts pointing to new partnerships being formed.

If **Jupiter** is the first planet in the first house, things will run smoothly; in such a year we can expect a relatively high proportion of luck in the unfolding of events or developments throughout the period. If placed in the second house, this will usually show up in the financial area (especially on the income side), but may also bring improvement in self-esteem or self-confidence, especially if the person is lacking in this area, or has recently suffered a setback, resulting in the loss of his or her self-confidence.

Let us not forget that Jupiter is the planet of expansion, protection and support, so generally a very good potential. If it is also well aspected with natal planets, the year in question is bound to bring a significant improvement in the previous condition or natal potential. Often we can even locate the area where improvement can be expected.

With **Saturn** as the leading planet there will be a lot to worry about. Things may become challenging, one will have to face responsibility, sobriety, stagnation, problems, lessons, maybe even illness (Saturn being the traditional ruler of health). Patience will be required. There may be only one major milestone in the year, however, it may later on also be referred to as the year "when nothing went right", with Saturn's inhibiting influence stretching over the whole year. (It is often difficult to delineate whether this is in fact a yearlong problem, because it is quite possible to be "stopped" by a single event and then feel its effects for a longer period of time, directly or indirectly.) Incidentally, this Saturn position is often also indicative of problems with authority, and it is not uncommon for a year like this to evoke power-battles with father, or for relations with him to deteriorate for some other reason. The loss of the parent, who had been authority for the person in question, is also not unexpected, such an event usually bringing an encounter with the independence we have not been used to before.

Uranus, as the leading planet, is no different than otherwise, i.e. bringing about reversals, surprises, innovations, novelties in approach, revolutionary or independence tendencies and the like, with its role here being even more noticeable and highlighted. Sudden and unexpected circumstances are possible, and some things will be completely different from before. This could be a year of upheaval or of strong outward change, or a year of increased independence. This is particularly the case when Uranus is placed in the second house as the leading planet, which can be interpreted as an attempt, or at least as a firm inner decision to become independent (at least financially, if not in terms of personality) from the environment to which the person (still) belongs at the time of the current solar return. Of course, according to the meaning of this position, the independence or the decision to become independent (which may only be possible at the first opportunity, perhaps not even in this solar year) may well turn out to be one of the key points of this solar return.

The planet of confusion, mystery and illusion, **Neptune**, as the leader, brings the confrontation with "something that is not the way it seems really". The illusion of self, the search for self, the wandering from one form to another, and back again, only to find something that we are sure is valid and real, but we don't realize that it is just a dream... Neptune is not likely to contribute in any way to sobriety and the achievement of tangible goals. Later we might even claim to have been lost, searching and perhaps even finding ourselves, but elsewhere than anticipated...

If Neptune has good aspects, it can indeed bring great inspiration for a work of art or the resolution of a previously unsolvable dilemma, but mostly it just brings confusion. But if Neptune has predominantly or entirely bad, unfavourable aspects, this "feeling of being lost" can manifest in much more concrete forms, such as alcohol, depression, paranoia. In such a year it is quite possible to realize that someone has been playing games with us, or that they have been taking advantage of our loyalty, willingness to help, or good humour in general, but we have never been aware of their abuse. This is also true of Neptune's position as the leading planet in the second house, as such a realization tends to diminish our own sense of worth and stability; for Neptune never gives a definitive answer as to whether or not we are really less worthy, for it is his job to keep us in suspense.

Yet the best explanation of this situation is that we have no idea why things are not going well, but in fact we do not even suffer and no harm is done.

Pluto as the leading planet brings financial problems or hardships (to do with shared or other people's money rather than loss of income), as well as some kind of reorganization or transformation. If we study the solar return of a person, whose natal chart has the Sun or Moon or stellium in the eighth house, we can safely predict that the potential for profound change will occur or begin in the very year, when Pluto is the leading planet in the solar return chart.

If well aspected (and if not also unfavourably placed in the natal chart), such Pluto can signify a major breakthrough, a birth of something (figuratively speaking), depending on the situation and the meaning in the horoscope (rulership). If it is a problem that has been pressurizing that person for a long time, perhaps for many years, and the pressure has increased to unbearable levels as a result, the year with Pluto so placed is just the right time to release those tensions and break through the dams.

A badly aspected Pluto is bound to result in some kind of destruction or suffering, sometimes in a Mars-like way, perhaps with fighting, blood, war; but even in a figurative meaning, such a year will never be easy and smooth. With Pluto less prominent we can expect deep insights, sometimes sexual themes, even encountering someone that will alter the course of our life.

V

TIMING

So far we have become familiar with the solar return horoscope, with the different possibilities of combining charts, as well as with individual elements of the chart, i.e. planets, signs, houses and other specifics like Ascendant, the leading planet, etc. The yearly horoscope shows the important things, making up the story of the time between two birthdays, however, the important question here is, **when** these things are bound to take place.

If we read various references on the interpretation of solar return in literature, or if we have our annual horoscope drawn up on one of the websites offering such analyses, we will be given a number of rather generalized interpretations, which will all be valid for the year in question, but lacking any indication whatsoever of the timing of the predicted event. Is it likely to unfold in summer, in May, or around the New Year...

IS „ONCE A YEAR" ENOUGH?

Let us look at an example from one such source:

"*Mercury indicates a year of full employment, with maximally activated mind. There is much learning to be done (...). Dialogue and exchange of information will be a constant presence. Venus in conjunction and Jupiter in square with solar Mercury allow for smooth communication and easy learning. The person will have a sense of being listened to by others. Great opportunities for travel.*

The double influence of Venus promises a basically pleasant year, abounding with harmony and love. (The few people in the person's life, aware of her situation, will be providing their full support, help and attention.) With Venus on the descendant, a significant love affair may be in the air."

Perception of events, relationships and environment will be discussed in detail further in the book, but for now let us just examine the "ever-present dialogue and exchange of information" in practice. Does the statement refer to a period in a year, or to several longer or shorter periods within a year, giving the person in question a sense of continuity, sufficiently pronounced to remember the year by a "constant exchange of data"? Or does the potential only point to an event sufficiently strong and significant (i.e. some extremely important exchange of information) to consider it important at the level of the entire year? How do we remember it? Will all the forecasts still be remembered at the end of the year? Not to mention the dilemma that immediately arises when two possibly relevant potentials unfold somewhere in the year - which one is 'the one'? ... What if some crucial and extremely important, perhaps even fatal, warning for our client were found by the astrologer - should the timing given be 'only' 'once within the year' - ?! And also: by giving generalized predictions, are we astrologers not getting closer to those we would like to distance from, i.e. various fortune tellers, "having such a nice way of saying things in general" ...? Astrology is an exact science and the information it provides should be as accurate as possible.

This is why individual authors use or look for other approaches, additional possibilities, to better place such projections in time and thus add usefulness and practical applicability to information. The formulation "once within the year" then turns into a more precise "in February" or "in three weeks' time", making astrological forecasting a completely different matter!

HOW IS THIS PERFORMED BY OTHER ASTROLOGERS

The prevailing belief that the solar return technique is not a predictive technique in its own right, and in particular that it alone cannot determine, when individual potentials of the year are to be activated and realized, has led to different approaches to the timing of potentials and activities within solar return horoscope. Similar to the choice of predictive techniques, some authors prefer to use one approach and others another, as far as timing in solar return chart is concerned. However, we will only make a brief survey of the most popular ones.

The first such method, cited by most authors of books on solar returns, is the use of the progressed Moon. The Moon travels about one degree of arc in a month, 12 degrees on average per year. It is therefore perfectly logical that during the course of solar return, progressed Moon forms many aspects to the planets and points the astrologer uses, including the house cusps. This means that even if only the basic aspects are taken into account, a whole range of data on activation of the individual elements of the solar horoscope within one year is available.

Another approach to timing the fulfilment of the potentials of solar return horoscope is the progression of the Sun through the year. The Sun is the only planet that passes in an even rhythm all the 360 degrees of the ecliptic in a year, reaching each of the planets on its way, while also forming many aspects.

The method is unreliable, essentially focusing on the transit of the Sun through the horoscope. In a similar way transits of other planets can be considered, turning the yearly solar return horoscope into a mere framework, within which the planets "work".

The third method is an expanded and upgraded previous approach, combining the solar chart with transiting planets. The transiting Sun "brings energy, thus activating every point it comes into contact with" during its orbit. The author of this method, Celeste Teal, also claims that these influences can be detected down to the day, due to the possibility that the Sun triggers an event when reaching a particular planet in conjunction, the activity being completed by the time the Sun forms a different aspect with the same body, e.g. a sextile, square or trine. The author attributes particular importance to the Sun's square and opposition to its own position, which, in simplified terms, could mean that the days exactly on both quarters and halfway through the year should be activated by the transit of the Sun. It is hard not to argue with this, as most people have to calculate their "half year return" or the date "6 months from their birthday". If these special dates really had any significance, or at least represented activated times with any frequency, these cycles would have no doubt been noticed and given special attention by people over the millennia, just as, for example, the tides, the seasons, and other phenomena and cycles, many of which are not observable by the eye, but can only be identified and determined by very careful and studious observation. It is completely impossible for such cycles not to have been observed by people over the millennia.

Celeste Teal is not alone in this, there are several "solar return interpretation manuals" on the web. One of the authors refers to timing with the following words: *"The transiting Sun moves at the rate of one house or one sign per month. Therefore, if you count the houses from the one following the house, in which the Sun is placed, onwards, the first house will be the first month after your birthday. The other months are counted in a similar way. Note the conjunctions of the Sun with the points and planets in solar return as they trigger events. Use all the inner planets to predict events, especially the Sun as the most important element of the solar return. Aspects formed by the Sun moving through the zodiac generate activity."*

In an annual horoscope, therefore, the Sun's influence is followed from, say, the sixth house onwards, triggering events from the seventh house, followed by the eighth, and in a predictable order through the year until the next birthday, when the same count will start running in the third house, followed by the fourth, the fifth, the sixth, and so on… The only question, then, is from which house to count, the areas apparently following each other in the same order month after month. There is no doubt that the Sun will activate the solar return planets differently each year, but if we use the birth horoscope as an aside, its planets will be activated by the same Sun on almost exactly ("almost" because of the leap years) the same day each year. If mankind has historically, through centuries of careful observation and recording of observations, detected and studied the zenith, the lunar phases, the overflowing of the Nile, the migrations of birds, weather and a million other phenomena, it is highly unlikely that it has never noticed that the same influence, depending on the movement of the Sun, has been at work on the same day of the calendar every year. Therefore, we have every right to be a little sceptical about this method of timing within the solar return.

The American author Raymond Merriman, a well-known financial astrologer, uses three approaches to date events within the solar return: progressed Moon in the solar return, transits, and progressed solar chart angles (about one degree per year). He considers this last technique, also discussed by the English authors Kirby and Stubbs, to be the most accurate among the three. The MC and the ascendant progress with the same speed as the Sun through the year, i.e. a bit less than one degree a day (here it is necessary to take into account the actual movement of the Sun, which does not move evenly throughout the year, so we have to work with ephemerides.) If, for example, the Sun progresses three signs and ten degrees through the solar return horoscope by a certain date, the MC and the Ascendant are also supposed to progress by the same arc. At these newly defined points they can form an aspect either to a solar return or to a natal planet. Conversely, if we are looking for the moment, when the so progressed MC or Ascendant (depending on what we are looking for) is making an aspect to a particular solar return or natal planet, we calculate the number of degrees from the starting position, i.e. the position in the solar return horoscope, add the resulting arc to the position of the Sun, and check the ephemeris for the day the Sun will reach that particular position. The method requires a little more calculating, however, I cannot claim it to be very reliable.

In addition to these, there have been other attempts to combine the solar return horoscope with other techniques, such as secondary progressions, transits and the like. The approach is perfectly logical, since even the basic principles of astrological prediction demand that, as a rule, indications of several techniques should be considered, roughly resonating with the theory "the more indications for an event in the different techniques, the more likely it is to unfold". This approach understandably requires additional calculations, for example combined charts, or at least inserting the planetary positions, obtained by the additional calculations, in the solar return horoscope.

In his book *"Transiti e rivoluzioni solari"* (*"Transits and Solar Returns"*) Italian astrologer Ciro Discepolo gives the following recommendation (rule of thumb), regarding the timing within the year of a solar return: *"In an attempt to best determine the time of the events (which is indeed a difficult task) brought about by a solar return, it is also a good idea to take into account the lunar return and the transits of the more or less fast planets."*

The approach of timing the solar return using lunar returns for the same year is also common, and will be discussed in a separate chapter in this book.

Perhaps the greatest progress in the research into how to pinpoint the precise "time of year" was made by the aforementioned James Eshelman. He divided the solar return year into half-years and quarter-years, and made special charts for the respective periods. Without going into too much detail, it can be claimed that the process is similar in idea to the method of breaking down the "solar" year into lunar returns - which can also be used to obtain more specific information on a particular date, and could be called "intra-year fine-tuning".

Bottom line: there is no shortage of approaches, methods and options, but one thing they all have in common, is combining in one way or another solar return with other prediction methods, always adhering to the principle that solar return is not a stand-alone predictive technique, and that it is not possible to predict time using this technique **alone**. This book, to the contrary, proves otherwise.

HOW WAS THIS PERFORMED BY DIANA

As already mentioned, my career as an astrologer began with the role of an assistant, mainly organizer and, when necessary, translator, for my friend Diana's astrological consultations. I soon noticed that she was not only good at timing by means of solar arc directions (or just "directions" in our case; she was simply unsurpassable at this technique), but also at interpreting solar return charts. All these charts were prepared by hand by myself; not only was I familiar with them, I had also studied several before the consultations themselves. I soon noticed the position of her forefinger during a particular "prediction", resting either in the beginning, middle or end of a particular house of the horoscope. After a while I was already confident enough about the pattern to ask her, if I was getting it right, or how the technique worked. Her simple explanation was of course just a confirmation of my assumptions. *"Look,"* she said, drawing into the conversation some of the charts we had been dealing with that day. In a few sentences she timed the positions within each house and linked them into a kind of a "system". *"It makes perfect sense that this planet represents an event halfway through her year,"* she said, holding up a client's chart, while I was frantically examining my notes of that conversation.

She explained this "logic" by the simple principle of moving in a counter-clockwise direction. All the planets transit and progress in this direction (ignoring the exception of retrogradation), whereas in the solar arc directions there is no such exception, this technique not recognizing retrogradation. The planets thus enter a sign - both transiting and progressed ones - at the house cusp and proceed towards the end of the house, i.e. the cusp of the next house. Whatever they activate in a house, they start with the planet closest to the cusp/beginning of the respective house, continuing to activate with the second and finally the third, if there are three planets in that house. In that simple order. In the hypothetical case of the first planet in the first five degrees of a house, the second in the middle and the third a few degrees before the next house cusp, all three would be activated in the same order. This means that even if there are no planets in a house, the beginning of the house comes first, followed by the middle and the end.

No one else in the same way?

At this point, it should be specifically emphasized that all my knowledge about the solar return technique originated from the few textbooks on the matter, so I was far from being familiar with it and even further from using it. This means that I had no doubts about the theories of the girl, who demonstrated exceptional practical knowledge, fascinating literally every one of our participants with her expertise. This is the way things are done.

No wonder I was surprised at the hypotheses that "timing is not possible in the solar return technique", that "it is a secondary and less important or even less reliable technique", and that "solar returns don't work", etc., abounding in the books I have come across. I was almost offended.

As a result, I started focusing on the dilemma of why things are the way they are. Why do astrologers claim that there is no such thing as timing, and why do I time the events, indicated by the positions of the planets in the yearly horoscopes, with such ease, and, in particular, why do I keep receiving so much confirmatory feedback? For in my preparations for consultations I, like most astrologers, have usually made at least one solar return chart for the year prior to the meeting, that is to say at least the last return that has already been completed by the time of the interview. My approach to this was usually to ask the client about an event, get confirmation and immediately proceed to the question **when it had happened**, with the proviso that I would often suggest the time myself.

The second group of these verifications includes the cases, where I myself enquired the client about the timing, while the third group comprises the cases where I focused only on a date and got feedback from a suggested period within that year on **what** should or did happen, and even in that case I often listed a few options that seemed possible or plausible to me.

There were so many confirmations that within a few years I gave up all doubt as to the efficiency of this simple method of Diana's. My subsequent search for confirmation, as well as the absence of any mention of this method in the available professional literature left me increasingly confused, so the aforementioned conversation with Dr Huber in Basel was a real balm for the soul. Not only did he make it clear that he was not familiar with the method ("if he as the most knowledgeable person does not know it", I said to myself, "then it really doesn't exist..."), he concluded by saying: *"If it works, then it exists, and needs further elaboration"*. I was speechless, but also relieved. From that moment onwards I have been more reassured, knowing that I have been on the right track.

„IN THE MIDDLE OF THE HOUSE"

In the simplest terms, Diana believed that *"the beginning of the house correlated with the beginning of the year, the middle of the house with the middle of the year, and the end of the house with the end of the year"*, this not being a calendar year, but a year between two birthdays, which we have already got used to. More specifically, a planet in a close conjunction with the house cusp points to the potential to manifest at the very beginning of the solar return year. The same is true of a planet, placed exactly in the midpoint of a particular house, signifying a potential to be fulfilled halfway through the year. ≈ to take place "in the quarter of the year" or "in the third", etc. A simplified graphic illustration of this principle can be seen in the figure below.

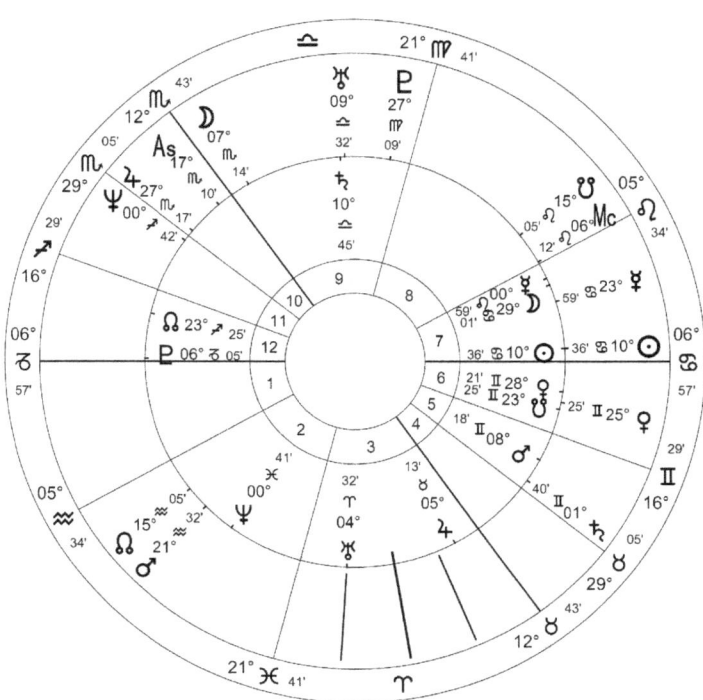

At this point it has already become apparent that any novice user of this technique will have to get used to counting degrees, subtracting, adding and dividing, all those not-so-popular components of astrology reminding us that the latter is largely about astronomy and (often unpleasant) mathematics. But not only that: since we are not going to calculate every little thing, having six pairs of houses in the horoscope, mostly differing in size, makes it advisable to be trained in another skill – making rough estimates. But even that will not be enough. If "year" in the solar return context means the time between two birthdays, a rough estimate will require a good idea of what "mid-year" is in each case. The sooner we have mastered rough estimating that the 'half year' for a person born on 18 October is 18 April, for example, the better. The same applies to the quarter of the year, i.e. 18 January and 18 July, as well as to the third of the year and other segments. It may seem a bit tedious at first, but it quickly turns into a routine... Even more, with a bit of practice, we will soon be able to determine the astrological houses, activated in 'mid-year', 'early in the year', etc.

CALCULATING THE MONTHS OF THE YEAR

If we decided to use the equal house system, the method would become much simpler. Each house would count 30°, meaning that half house would be at 15°, etc., but also that a segment of 2.5° would represent a month - like one twelfth of the year. One would then just have to count the months... Ideally, the Ascendant as the starting point of such a chart would be at 20°, 10° or even at 0° of a sign, and counting would in that case be even easier, all the houses also starting at 0°...

But this is a rare luck. In reality, most astrologers do not use the equal house system, but other systems. There, each pair of opposite houses has to be dealt with separately. Counting the months with the Ascendant at, say, 13°47', becomes a little more uncomfortable, what with the seven months we are supposed to estimate, calculate...

Depending on the latitude and the Ascendant/zenith (or a system of houses using one of these points as a starting point), the calculations can find astrological houses (or pairs of houses, since we know that opposite houses are the same size) of different widths. For example, for a latitude of 45°, where Slovenia is located, the range of the astrological houses varies between 18 and 62° [5].

This roughly means that in the first case, a particular month will only have a 1.5° division within the astrological house, and in the second case, a little more than 5°!

The most straightforward examples for calculation and assessment are:
- house size 18° - 1.5° per month
- house size 24° - 2° per month
- house size 30° - 2.5° per month
- house size 36° - 3° per month
- house size 42° - 3.5° per month
- house size 48° - 4° per month
- house size 54° - 4.5° per month
- house size 60° - 5° per month

---

5   With some practice in astrological computer programmes, we can see that the astrological houses differ less and less in extent as we approach the equator, because the Midheaven (MC) also moves less and less away from the zenith, while at higher latitudes the MC tends more and more towards the Ascendant or Descendant as the day progresses, to the extent that the houses of a quadrant are sometimes compressed to just a few degrees.

Intermediate values are a matter of estimation, which, like thousand other routine tasks, is a matter of practice and mileage.

For several months - 2, 3, 4 - it is the estimation or calculation of the proportion of time within a year that comes into play, rather than multiplying the values for each month and multiplying by the corresponding number. This way we quickly learn how to calculate a third of 57° or 28° - a much quicker and therefore more useful process than dividing a house into twelfths and then multiplying the number by months, the possibility of calculation error being smaller with this method. To reiterate, this is only a more or less rough estimate of the division of houses into time-appropriate segments; a few computational operations are nevertheless necessary for a more precise value. However, with years of experience, one reaches the point where calculation is less needed.

For those less skilled in mathematics that find dividing houses into twelve equal parts too troublesome, the handy table in Appendix 2 will be useful.

Shock on a business trip

A few years ago, while preparing materials for a presentation of the solar return technique, the case of a client's spouse, who had recently suffered an extremely unpleasant experience, came across my desk. The solar return chart showing this event is presented.

As the gentleman was born on 22 October, this will also be the starting point for our calculations. On Friday, 11 March 2005, he was returning from a business trip in a foreign country. It had been an extremely successful trip - as a salesman he had signed lucrative contracts one could only wish for in this business. Expecting to win praise for the contracts he was on his way home to his family... When the phone rang, he answered it without hesitation, happy to be able to tell his boss the good news straight away. But the shock couldn't have been greater - his boss told him, very briefly, that he had been fired and that he was only to report to the company on Monday to tidy up his desk and pick up his employment record book?! Like in a badly directed film, he was sitting in his parked car at the side of a foreign road for a while, staring at his phone...

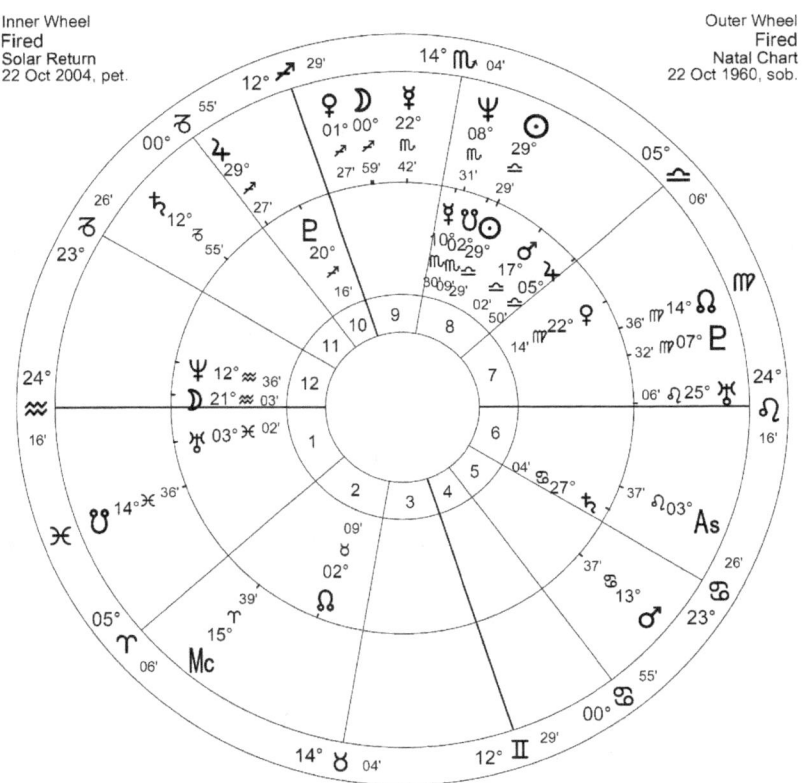

What could be the astrological meaning of a "shocking job loss"? Perhaps Pluto (destruction) in the 10th house (career) could be the answer. And when we look at the man's solar return, Pluto is indeed right there! A complete shock in the work field. But let us take a more specific look, using some mathematics.

NOTHING WITHOUT RECALCULATION ...

The tenth house in this horoscope begins at 12°29 Sagittarius and ends at 0°55 Capricorn. Its magnitude is 17°31 degrees within Sagittarius and another 0°55 in Capricorn, making a total of 17 degrees and 86 minutes, the same as 18 degrees and 26 minutes.

(An alternative way of calculating this, which may be used in more complicated cases, is as follows:

Since the end of the house is already in the next sign, add 30° and then subtract the first value, the beginning of the house. So: 30°55 - 12°29 = 18°26.)

To divide into 12 months, we first need to convert the range in degrees into minutes to make division easier:

18 degrees and 26 minutes = (18 x 60 minutes) + 26 minutes = 1080 + 26 minutes = 1106 minutes

1106 minutes : 12 months = 92.17 minutes/month = 1 degree and approximately 32 minutes for the month

From here on, several options are available. Because we can see at a glance that Pluto is positioned about halfway through the tenth house, but still a little earlier, the event probably took place sometime in March. We will therefore focus on the time about a month or a month and a half before the half year, i.e., 22 April. Halfway point of the tenth house is 9°13', and added to the beginning of the house, this is 21°42 Sagittarius.

Pluto in this house is placed at 20°16 Sagittarius, which is 1°26 away from the midpoint of the solar year, i.e. a little less than one month before the midpoint. This brings us to a few days after 22 March, perhaps March 24 or 25.

Alternatively, we can simply calculate the fraction of time between the start of the solar year and the time indicated by Pluto's position, and convert this into days. The difference between the house cusp [6] and its position is exactly 7°47, which amounts to 467 minutes.

467 : 1106 = 0.422 years

0.422 x 365 days in a year = 154.03 days, converted into months and days this makes 5 months and 4.03 days

Adding this to the date of birth gives 22 March (5 months) + 4 days = 26 March

However, the following calculation is also possible:

The difference between the house cusp and Pluto's position is exactly 7°47, amounting to 467 minutes. To get the number of months (and days) from the birthday to the fateful phone call, divide this sum by the value, obtained for one month in that year:

467 : 92.17 = 5.07 months

The remainder of the 0.07 months is converted into days, i.e. multiplied by the number of days in the month [7], which is 0.07 x 30 = 2.1 days, which together with the 5 months from the first part of the calculation gives 24 or 25 March.

And how were things like in reality

What have we learned so far? That it is not true? The three results give a time frame within one or at most two days, which is encouraging.

---

[6] If the planetary position is close to the end of the house, it is logical and easier to calculate the difference between Pluto's position and the end of the house.
[7] To be more precise or consistent, we should take 30.416 days as the number of days in an average month, since not all months are equal, and in a normal year there are 5 days more than 360, which is 'a multiple of 30 days in a month'. For a leap year, we would then have to take 30.5 days, because there are 6 extra days.)
(The calculations produce very similar or identical results, which is also affected by the conversion between units, as well as rounding up of the resulting values.)

We now know how to calculate and estimate the time distribution in many ways, but this documented example raises the old question: what is it that we actually see in the horoscope? The call was made on 11 March, after all, and not two weeks later?! But we also wonder what happened after the call. Let us take a methodical approach.

The discussion of this case took place at a distance of time, a few months later, the reason for the visit to the astrologer having primarily been the start of a new private business as a result of the job loss. We actually calculated a two-week delay, since the "fateful call" took place on 11 March. If the man had been born at 22.57, the Ascendant, together with the house cusps, would have been slightly shifted, and our calculations would have indicated 4 months and three quarters after his birthday, i.e. 15 March, which would have been quite accurate. But the birth time '22.55' is demonstrably correct. What is this Pluto pointing to then?

Had we discussed the affair when it actually happened, with emotions and impressions still fresh, we could have gained a better insight into the actual events and perceptions. We know that, although Monday was a turbulent day, the man still expected things to be sorted out. Pluto being the ruler of the 9th house in this chart shows that he tried an appeal or other legal action, but nevertheless learned quite quickly that it was not going to work and that 'ships were sinking'. And then a few days passed - that inner, intimate realization that 'nothing more can be done' did not come on 11 March but later.

Does Pluto point to this? The planet certainly means "destruction" (or at least a feeling of destruction, fear, helplessness...), the 10th house means "career", and this turning point brought a purely Plutonian quality - the death of something and the birth of something new, not least because it forced the man to move his business into a completely different field, in which he had never worked before.

And one more thing to remember: from the distance of time, only the most solid records and memories remain in our perception and memory, and we know from practice that sometimes "things come after us".

A FEW MORE CASES

A diploma and a new job

Another very illustrative example, which simply has to be included in this book, started - after several consultations – with a telephone message about the client's success, her graduation. As she did not suggest another 'look at the horoscope', I kindly congratulated her and never bothered to look at the calculations. Two days later, however, she called again, asking about the offer she had just received, 'which she was not to refuse, but still had mixed feelings about'. The offer had come from the very institution, where she had recently done a traineeship for a few months as part of her final studies.

The chart of the solar return horoscope, corresponding to our last consultation, was still in the archives, including all the typical notes, lines, numbers, question marks and the like. Thus, by the position of the Sun in the tenth house, there was an "offer h." note, meaning that we had apparently already mentioned the offer from the hospital as an option, probably at the time of the conversation, and close by there was also a note "j. 02.17", referring to a "job in February 2017"; the consultation having taken place at the end of December 2016. With the Moon's Node in the ninth house (studies), I had marked "fac N.Y./01.17?" at the time, suggesting that we were discussing the option that somehow the best indication for that particular time was that she could complete her studies. The opposition of the Node to Neptune in the 3rd house at this time suggests the possibility of some unforeseen situation or trouble. Two factors, however, were best suited to the term of the degree: the sextile and trine with the pair of **natal** Moon's nodes, which natally are placed at the very edge of the 3rd and 9th houses, along with the position of the solar Mercury, ruler of Virgo, in which this node is placed; Mercury is exactly on the MC, thus connecting the 9th house of study and the 10th house of achievement.

Graduation is certainly a theme, related to the ninth house, but not necessarily. Many astrological sources and authorities associate its indications with pairs of signs or houses, which means that sometimes the indications of study or graduation are also found in the opposite, i.e. third house. But things get even more complicated, an individual's perception of graduation (not of studies in progress!) often being linked to concepts such as achievement, success, triumph, realization or status, all of which are attributes of the tenth house. It is therefore not uncommon to find clues about graduation in the tenth astrological house.

All this, of course, immediately aroused the astrologer's investigative streak – let us see how accurate this can be!

The 9th house is extremely large in this solar chart, between 26°40 Leo and 18°28 Libra, a total of almost 52 degrees. This means that 1/12th of the house falls per month, which is about 4 1/3 degrees. The client's birthday is on 19 October, so the months in this solar year follow each other in degrees as follows:

1° Virgo is 19 November, then by month: 5°20, 9°40, 14°, 18°20, 22°40 (mid-year, i.e. 19 April), 27° Libra, 1°20 Scorpio, 5°40, 10°, 14°20 and 18°40. However, since the house actually ends at 18°28 rather than 18°40, we arrive at a difference of 12', which we divide by 1' for each month to get a more accurate monthly breakdown:

0°59 Virgo, 5°18, 9°37, 13°56, 18°15, 22°34 (half of a house), 26°53 Libra, 1°12 Scorpio, 5°31, 9°50, 14°09 and 18°28.

The month of observation is February, i.e. the division between 9°37 (19 January) and 13°56 Virgo (19 February). The Moon's node is at 11°53, i.e. 2°16 from the first position and 2°03 from the second. Dividing 4°20 into the 30 days of the month gives about 8.355' for a day, meaning that the position of the Moon's nodes points exactly to 4 February, which was a Saturday that year. She graduated on 6 February and posed the question about her job on 9 February. A difference of just one minute in the time of birth could have brought the accuracy to within a day…

And the outcome? The aforementioned opposition to Neptune actually occurred at a time that could be calculated with similar precision: about two weeks before graduation, the professor suddenly ended up in hospital for observation, and a few days later she returned to work, without specifying the kind of collapse she had suffered from (a typical Neptunian influence when health-related).

What to do with the two days difference? Preferably nothing, because a difference of just half a minute in the time of birth (which no one knows for sure, because, for example, the wall clock in the maternity ward could have been one minute behind or one minute ahead…) would give us changed calculations and, as a result, most probably a very accurate day. However, the question here is, whether such a precise calculation makes sense. In some cases yes, in most cases not. Anyway, our client used this calculation to schedule her commitments and accomplished them according to the plan.

Also interesting in this context is the neighbouring 10th house. There, the Sun is in a similar position timewise. The calculation gives us its exact position as "2 days more than 4 months after the birthday", taking into account the division of the year shown above. This brings the date to 21 February. Although she was unable to confirm this at her next visit, which took place a year or so later, the truth was that she "got the job in February"...

Too complicated? Probably yes, at the first glance, but in fact it is not so bad. Practice brings skill and shortens the time; moreover the ability to assess values proves extremely useful here. Even in the course of writing the present text, it turned out to be much more difficult to formulate these mental steps than to carry them out in practice – the aforementioned calculation took only a few minutes to produce the date, certainly much less than the time needed to check and write it all down. So go ahead without fear ...

A life turning point

The loss of a loved one is one of the greatest shocks in a person's life. It happened to Miss Majda on 13 June 2006, when her husband suddenly died. But it was not only the loss, there was also the impact of this death upon her relations with her surroundings, not to mention the material situation – in short, everything was suddenly turned upside down.

The solar return horoscope of this case is most eloquent. The alignment of the two axes of the lunar nodes, natal and solar, with the horizontal axis and with the Sun points to a year of great importance. Saturn is at a culminating position, in the tenth house of its rulership, speaking of stagnation, problems and possibly fear. The two Malefics, Pluto and Saturn, the latter as the ruler of the natal Ascendant, are placed in the eleventh (future) and twelfth houses (depression, helplessness). The sixth house is full. If the aspects were marked in the combined horoscope, one could easily notice a very exact Grand Cross in mutable signs, consisting of the Moon,

Mercury, Mars and Saturn, along with a T-square in fixed signs, consisting of the two Venuses, Neptune, Pluto and Jupiter as the focus of the configuration. Moreover, the two configurations are at very similar degrees, so they are closely intertwined.

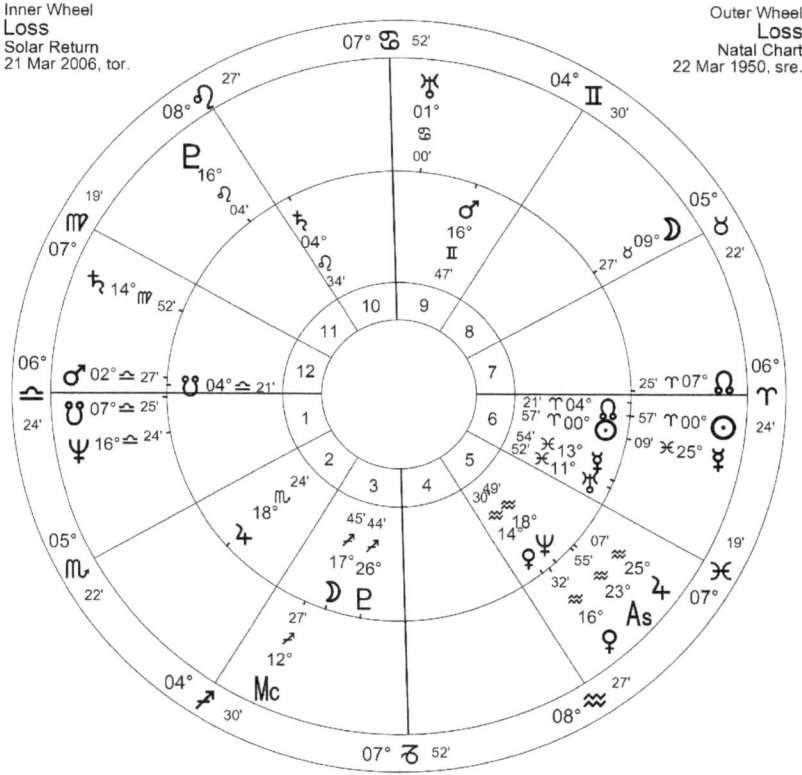

In this case, 13 June comes 81 days after the date of birth, so the event could be visible in the horoscope at a point just before a quarter of a solar year (81 days = 0.222 years). The relevant degrees in the astrological houses here are occupied by Mercury as ruler of the 12th house, i.e. of the husband's 6th house, denoting his health, and the two Venuses, the solar one as ruler of the ascendant and the natal one as ruler of the 8th house. (Let us not overlook - in her natal chart Saturn as the overall ruler of health is placed in the 7th house (spouse), inconjuncting Venus in the 12th house, also ruler of his 8th house; that makes her husband's health an important area of her life.)

The Grand Cross is timed for later in the year, a little before the middle of the year, August or even September. Both configurations are activated at the time, denoting lots of activity. (The planets in the configurations are not at the same degrees in their signs, but they are all within four degrees, which, together with orbs and midpoints, if taken into account and calculated, suggests a longer period within the year, say a few months.)

In this case, things got worse within a month or two, the challenges and problems having been so overwhelming, that the client did not know what to deal with first. Pluto - together with Saturn and the fear and helplessness it brought into her life – made her fear the future, Mars in the ninth house brought about great demands and paperwork procedures, and the Moon (together with Pluto a few months later) generated changes and then a conflict with her family (third house). Neptune in the first house made her question her abilities and strength (never a small thing for an Aries), while solar Neptune and natal Ascendant in the fifth house generated a feeling of not being loved by anyone. (Natal Uranus as the ruler of the natal Ascendant is adversely placed right there in the fifth house, bringing the potential for the client to stand up for and love herself more, which certainly came true this very year.) The strong focus on the sixth house here is not a significator of health issues, but rather refers to tidying up and creating a new order in her life.

Further examples, specific features and curiosities will be discussed in the next chapter.

VI
STARGAZER'S COOKBOOK

We have so far become familiar with the solar return technique and acquired lots of useful information on its options, and yet new questions, doubts or borderline situations keep coming up. What to do with retrograde planets, how to deal with configurations, how to explain "empty" houses, how to make a meaningful whole, and so on. In the thirty years of astrological work, a wealth of experience, insights, knowledge, shortcuts and useful tricks has been accumulated, so the astrologer's notebook is full of records, notes, etc. An attempt to organize the material brings about a cookbook, full of recipes and hints on how to prepare a dish. It is impossible to write everything down, so let us only deal with the most common dilemmas and questions.

ASTROLOGICAL HOUSES

In the previous two chapters we have already come across the houses, regarding calculating the proportions, as well as their interpretation in solar return chart. The knowledge of the astrological houses being part of the "iron repertoire" in educational programmes, reader will certainly be familiar with most of the information required to use them in the horoscope, however, it makes sense to refresh some of the facts, or even to discuss some not so often encountered views.

Koch or Placidus

The question whether an astrologer uses the Placidus or the Koch house system in his calculations usually turns up at the beginning of a course. The dilemma of which of the systems is "better", or at least more useful, is a constant in astrology, never to be resolved probably. Moreover, it is difficult to estimate the number of astrologers using one or the other, everyone having his own argumentation as to the choice.

I myself entered astrology with the Koch system of houses and have been using it for a number of years. But a researcher's mind never being at rest, and the question of 'this or that system' – there being many more, of course – coming up incessantly, I decided, at some stage of my professional career, to resolve the dilemma "once and for all". For about half a year I was making most of the calculations, both of natal charts and of yearly horoscopes, in parallel, so that I had two decks of charts for each consultation! Quite a time-consuming work, but with a clear intention to **practically** find out, which of the house systems suited my work better.

Every beginner astrologer knows that the vertical axis (MC/IC) tilts alternately sharply to the left and to the right every day, reaching its maximum deviation from the vertical with Ascendants at about 15 degrees of Gemini and 15 degrees of Capricorn, making some houses very narrow and others much wider. But as the two systems are identical in the angular points and different in the intermediate houses, i.e. in the cusps of the 2nd, 3rd, 5th and 6th houses, the same being true of the cusps of the four opposite houses, the position of a planet in a particular house may vary a lot in different systems. If we have mastered the technique of timing the position of a planet in a house, we may arrive at a completely different result within the solar year with a different house system. Moreover, it is not uncommon for a planet to change houses and no longer be in the same house, but in an adjacent one. An example can be seen in the figure.

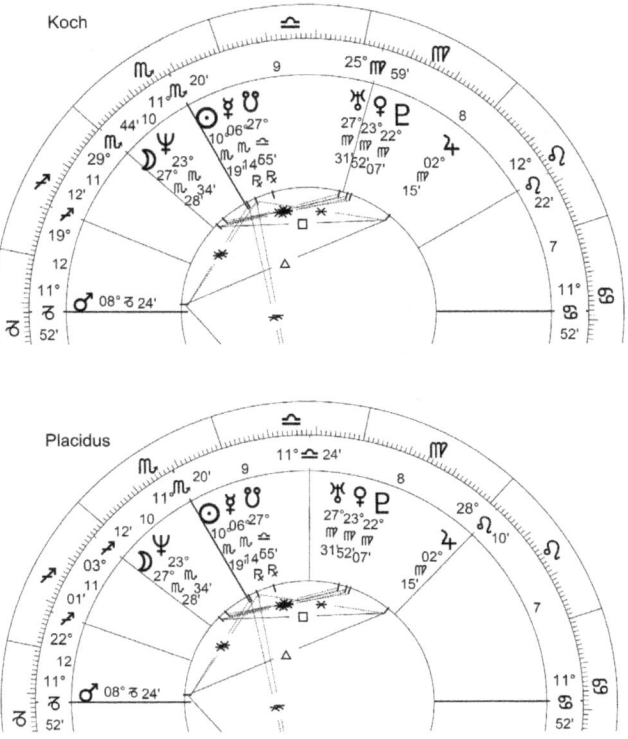

With such a colourful array of information acquired, it was not difficult to work out the pairs of horoscopes corresponding to the respective stories or events. I have henceforth only been using the Koch house system, which I do not claim to be superior; in the course of time it has just proved useful to me and my technique. Every astrologer, of course, is bound to find sooner or later a system of houses that suits him best.

Northern and southern latitudes

A different problem arises with horoscopes of people born in the higher northern latitudes, including, for example, Scandinavia, much of Russia and Canada, and a few other countries. With these regions, horoscopes, based on Koch and Placidus, made for certain hours within day, turn out rather unusual. Namely, an entire quadrant of houses, in this case the first or the opposite third, is squeezed into only about 15 degrees of arc, making any timing of the planets in a solar return meaningless. The same is true of natal horoscope itself, so the rule stating that these two house systems are only useful up to 66.67° north and south latitude, i.e. up to the two tropics, is quite relevant. In practice, this proves to be inconvenient even with horoscopes of people living in the UK, Scandinavia or Iceland at latitudes higher than 50°.

In the few cases of having worked with people from Finland or Scotland, I have been advised by a fellow astrologer, who lives and works there, to try the Meridian house system. She does not use solar returns herself, preferring other predictive methods, however, this system has proved to be very accurate with the northern latitudes.

Three signs in the same house

Due to the peculiarity described above, where the Midheaven (MC) oscillates to the right or left, some houses in the horoscope happen to be very wide. Knowing that this range can amount up to 62° in latitudes around 45°, such houses will obviously extend over three zodiacal signs.

We are already familiar with the basic explanation of this phenomenon from natal astrology, but for solar returns this possibility provides more information. The ruler of such a three-sign house is still the planet ruling the sign at the house cusp, regardless of where the cusp falls - at the beginning, in the middle or at 29°59 of that sign. In the presented example, Mercury rules Virgo on the ninth house cusp of the horoscope. However, this wide house embraces the whole of Libra and finally a few degrees of Scorpio. Venus, Mars and Pluto, respectively, are thus the co-rulers of this house, not ruling the entire house, but only the parts they occupy. Venus rules the major part of the house, Mars and Pluto only the minor part at the end. Let us now examine such a solar return from the archives:

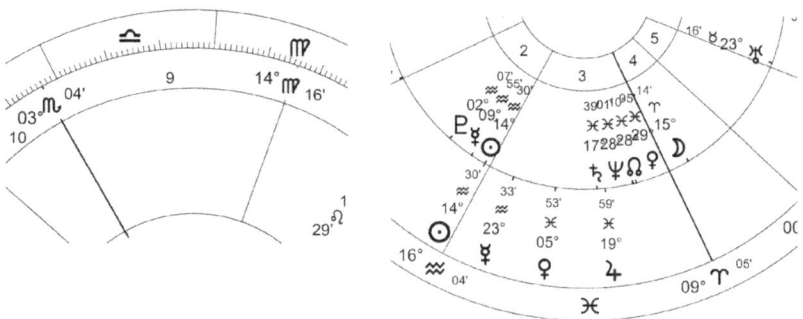

This is the third house, starting at 16°04 Aquarius and extending through Pisces to 9°05 Aries. There are three natal and solar planets, including a solar lunar node, a total of seven elements. There are also two house cusps, which together provide intense activity in this area, along with numerous data for the astrologer's interpretation.

An interpretation of this astrological house – without considering all the aspects and rulerships – would be as follows:

"The house starts in Aquarius at 16°04, with Mercury retrograde also in Aquarius at 23°33. Then the house moves into Pisces with natal Venus at 5°53 Pisces. Then somewhere in the middle of the sign, Saturn (at 17°39) and natal Jupiter (at 19°59 Pisces) are placed close together. There is a lot of activity at the very end of the sign, with solar Neptune at 28°01, conjunct the North Node at 28°10 and Venus at 29°05 Pisces. The house then moves into Aries, with no more planets there until the end of the house at 9°06 of that sign."

The house itself can be divided by months as follows:

16°04 Aquarius, the beginning on the birthday itself, i.e. 3 February of the year; followed by: 20°29; 24°54; 29°19 Aquarius; 3°44 Pisces; 8°19; 12°34; 16°59; 21°24; 25°49 Pisces; 0°14 Aries; 4°39 and then 9°05 Aries on the next birthday, 3 February.

Accordingly, we can more easily determine the parts of the year when each element is active, Mercury about two months after the birthday, with the house then moving into Pisces sometime in the first half of May this year. Venus is expected to be most active sometime in mid-June, and Saturn and Jupiter in the first half of September, just after the midpoint of the solar year. The most intense month of the year is obviously November, with Neptune, the North Node and Venus in a tight stellium there; their conjunctions and rulerships point to the areas involved, together with the person's natal potentials. The period also coincides with the passage of the house into Aries.

The transition of an astrological house from one sign to the next is also a point to be felt during the year, not so much through events themselves (or very rarely), but through a change in the perception of the mood or atmosphere of the respective house. If the third house in this example were observed through relationships with relatives (it could also relate to short journeys, traffic, etc.), we could anticipate distance or defiance in these relationships at the beginning of the year, whereby Mercury (retrograde = going back or to old positions) could denote an initiative "from outside", i.e. not from the respective person. A move into Pisces could refer to disbelief, perhaps a bit of idealization, regarding outcome. Venus then – depending mostly on the role and placement in the natal horoscope – refers to benefit from others, which this person accepts or uses without straining the relationship. Saturn could bring some stagnation, fear, frustration or inability to change into the mix, while natal Jupiter immediately afterwards, within a month or less, brings "solution" or offer from the other side, along with a warming up or relief. The intense activity in mid-November moves from disappointment (Neptune – "they didn't have honest intentions after all!") through an inner decision how to act henceforth, and to Venus, which could suggest a new attitude, with a good deal of optimism and goodwill, but also calculation. The move into Aries here means that by the end of the year, the person will be conducting these relationships in accordance with her own intentions, without any regard to the attitudes or expectations of others, perhaps - depending on the position of Mars in this solar return - even more harshly or exactingly. Further developments will be evident in the third house of the next solar return, representing a logical sequel to the story...

An empty astrological house can be "read" in a similar way - although it may not contain planets, it may, for example, contain an eclipse or a fixed star (if used in astrological work), but anyway the rulers themselves and their conjunctions provide several possibilities. Also relevant may be squares and oppositions of a planet, placed elsewhere in the horoscope.

To sum up: basically, it can be claimed that a wide astrological house means the respective area - and of course the area covered by the opposite and equally wide house - will be very important in the current year. In this regard, it will sometimes be necessary to decide, which of the pair of houses is the one most highlighted, e.g. the second or the eighth.

Similarly, the narrow house is not considered to be very important in the year, although that does not denote any lack of activity in the respective area. It is not the size of the house that matters, but the number of elements it contains – all in accordance with the principle of "the more planets in the house, the more dynamics". Which means that a very wide house could be empty, denoting a highly important area with little dynamism, certainly less than we would have liked, or conversely, a narrow house could contain several planets, denoting intense activity in an area we are not particularly interested in, at least not in the current year.

House rulers in solar return horoscope

Familiarity with rulerships in astrology is a fundamental part of basic astrological education, so the system itself is not discussed here. However, since the related terms have frequently been used in discussing the method and interpreting cases, and due to the likelihood of incorrect assessments or connections in the course of our work, another distinction has to be pointed out.

As there are two circles in the solar zodiac, the solar and the natal one, there are twenty planets, ten in each circle. Each horoscope has its own set of houses, beginning at the natal Ascendant and the solar Ascendant, respectively, and these houses are rarely ever in the same signs or close to one another. The planets therefore rule the houses of the pertaining horoscope - natal planets being rulers of natal houses and solar planets being rulers of solar houses.

In this way a whole lot of new highlights, planetary strengths and roles is acquired. Mercury or Venus (and of course Mars, Jupiter and Saturn when considered traditional rulers) thus rule two houses in a horoscope; with stronger tilts of the MC/IC axis they can even rule several, including two or more in a solar horoscope. A planet in the natal chart may only rule a minor house, while suddenly acquiring a major role in the solar return – as the ruler of the Ascendant.

However, it should not be overlooked that the newly discovered planets, i.e. the trans-saturnians Uranus, Neptune and Pluto, rule three signs, which also have their "old", traditional rulers. The question, or rather another 'eternal' dilemma in astrology, concerning the more relevant ruler of e.g. Aquarius, Saturn or Uranus, is beyond the scope of this book, and a matter of reader's decision.

Overlapping house cusps

Cases of planets being placed very close to a house cusp, even by a single minute or less, are not infrequent. These planets deserve special attention in interpretation and especially as far as timing is concerned. This may sound illogical, since the position of 0°00 relative to the house cusp is the easiest to determine, falling 'exactly on the birthday'. But it is not quite that simple. Just a minute or even half a minute at the time of birth - and no one bothers to register half a minute - can bring a slight shift in the position of an astrological house, which in turn means that this planet can be placed at the very beginning of this house or at the very end of the **previous one**. This is not the same thing, of course. It is therefore worth making a pause in such a case and do some additional checking of the natal chart.

In this context, however, one more peculiarity should be mentioned, namely **an event** precisely on the occasion of the birthday. In my yearlong work I have come across cases, where an event corresponding exactly to a birthday, was not represented by any planet in the horoscope, placed exactly on the cusp of the pertaining house! As if this record had been hidden somewhere! That is why it is also necessary and welcome to track records of events in other prediction techniques...

STAR TRIVIA

No birthtime available

When working with a natal chart, an astrologer has a lot of options to work with, even if the birth time is not known. Even without chart rectification, for which several procedures that astrologers have developed over the centuries for their own use (more on this later in this chapter!), are available, he can resort to other interpretation options, including the equal and whole sign house system, etc. I use a method known in the astrology software as 'Sun on 1st', which is useful enough to save the trouble of having to go through lengthy rectification procedures.

A solar return can be made from any astrological chart, but the question is, whether such an annual horoscope can be of any use. Notwithstanding the fact that some authors work a lot with returns, one of them being Anthony Louis (who uses precessed solar returns), I myself have not found solar returns useful in such cases in many years of examination, and prefer to use other methods in my predictions.

To illustrate: as with natal astrology, the sword of Damocles hangs over the astrologer's head in the form of "what if the exact birth time comes later" question. For if we try to determine the periods of the year in a solar return from a natal horoscope without the birth time – regardless of whether we take '0.00' or '12.00' as a starting point - there is a chance, but only just a chance, that once we have a complete chart and the Ascendant and houses available, the highlights will overlap with those predicted earlier. To a certain extent, the impact of the power of thinking - or the placebo effect, if you prefer ... – may be at work here.

Natal and Solar Ascendant

Along with the twenty planets in the two circles of the combined solar return chart, there are also the two Ascendants, the MCs and other points, depending on the selected set used in the work, be it Chiron, Lilith, Part of Fortune, any set of fixed stars, vertex, etc. Each of these points has a position in the natal horoscope and a different one in the solar return, except of course the Sun. All this provides a variety of information, increasing the possibilities of a more adequate interpretation.

One highlight that has not yet been mentioned, and is by no means negligible, is the position of **the solar** Ascendant in the natal chart. It is obtained either by marking the solar ascendant in the natal chart, for example on the circumference, or by using the "Swap" option in the computer program (e.g. SolarFire), which turns the horoscope into a mode with the natal horoscope inside the combined chart.

This position will give us another indication of the highlights of the year. But it will get even more interesting if a series of solar Ascendants ("stamps") is inserted into the natal horoscope, showing how these highlights line up in the sequence of years...

Repeated positions

As we have seen, due to leap years, the 'stamps' of the solar returns (i.e. the annual Ascendants) line up in a complex order, each year a few signs further ahead in the zodiac. The natal planets therefore fall in different astrological houses each time, occasionally even in the same house as in the natal horoscope. But the planets move forward in their cycles throughout the time since our birth at their own speed, which means most of them also end up in the sign they occupied when we were born. But even so, a solar planet sometimes - albeit in a completely different sign - finds itself in the same house as in the natal chart. Then such a planet, although with a different energy, resonating with that of the sign of the solar horoscope, works strongly in the same area as in the natal chart. For example, when Mars, natally placed in the tenth house, finds itself in that same house of a solar return, the person is bound to, in accordance with the natal premise, make some important step or breakthrough in his or her business, career or other path in that very year, denoting success, victory, recognition or whatever else is associated with the tenth house.

Moreover, it is very common for a potential indicated in natal horoscope, but failing to unfold, to be activated in the very year when the respective planet is placed in the same house as in the natal chart. What is more: the aspects of this planet with both, the solar planets (current potential) and the natal planets, provide ample information on the progress of the respective potential, on the key factors, participants, etc.

To some extent, the status of the 'repeated planet' also applies in cases where, in a solar return, a planet is not in the house of the natal horoscope, but in a sign analogous to that house. This means that a natal Jupiter in the second house can be very strong, acting as the key factor in a solar return, even when not placed in the second house but in Taurus. And the other way round - e.g. Mercury, natally placed in the seventh house, finding itself in Libra in a solar chart.

This is not the end of the set of 'repeated' positions. A planet may form the same or at least very similar aspects in the solar return as in the natal horoscope. For example, a fast planet is often conjunct a slow planet, and in a given year their distance may be exactly the same as in the natal horoscope. (Such an example is given further in the book, in the chapter on relocation.) This applies not only to conjunctions, but to all possible relationships, aspects, conjunctions and configurations.

Eclipses

Eclipses are important triggers in life. There are usually four eclipses a year, and every six months both solar and lunar eclipses fall in opposite parts of the solar chart, usually also in the opposite houses. If we chart the close eclipses, i.e. the last pair before the birthday and the pair occurring half a year later, i.e. at the time of the current solar return, and also the pair after that, we can follow their movement and ascertain the impact of a series of eclipses upon a person's life.

SOLAR RETURNS

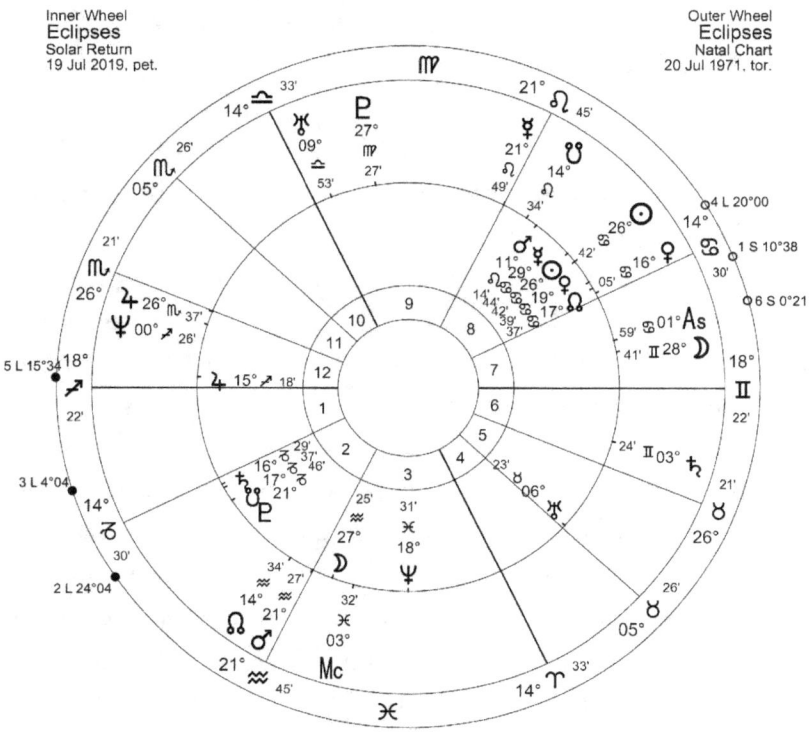

Different astrologers advocate different approaches to dealing with eclipses, especially in terms of the orbs taken into account, and the range of eclipse aspects, but even if we only consider the two basic aspects, conjunction and opposition to planets and angular points, not going beyond a 5° orb, solar horoscopes where these six eclipses do not touch at least one planet or angular point, thus activating them, are rare to find. Then, after trying to ascertain how much the eclipses' activation fits in with the natal premise, a particular solar return may turn out to be much more significant than we had assumed...

Asterisks in a solar return chart

Once we can determine (admittedly only more or less approximately) the times of the year, when planets are activated in a solar return horoscope, relatively quickly and without much calculation - not forgetting that moves to the next sign within a house also resemble a planet being placed there! - we will notice that some of these planets are within their houses at roughly the same distance in time from the cusp. This indicates roughly similar timestamps in the current year - the planets will be active at roughly the same time! Let us mark them with asterisks (or some other arbitrary and identical signs), check them by degrees and determine the deviations. Considering the orbs of action, it is possible that planets, predicted to be active in, say, May and June, may in fact be "active" at the same time. One may have already been active for some time, when the other one joins it, etc. In the course of my practice I have also come across cases, where seven planets were active over a short period of time in the year, marked by seven asterisk signs...

One such example is shown in the figure. No less than five planets are placed in almost identical positions within their house, i.e. halfway through, but Venus, to be very precise, is placed slightly earlier.

But there is a catch here: these planets are mostly not aspected to one another! So basically, they cannot interact at all! But that is not the case. Even if different areas of our lives are active at the same time, these planets are mutually interrelated in terms of **time** and **meaning**, not aspect.

A perfect example of such interrelatedness are temporally linked 10th, 6th and 2nd houses, denoting career or job, work or workplace and earnings. A common example of this phenomenon in solar returns is also the notion of relocation, where the 3rd, 4th and 6th houses are usually activated, with planets placed in similar proportions within the house, say three months after the birthday in each house, which certainly indicates intense activity in the respective life sphere; astrologically it can no doubt be inferred that relocation will indeed take place.

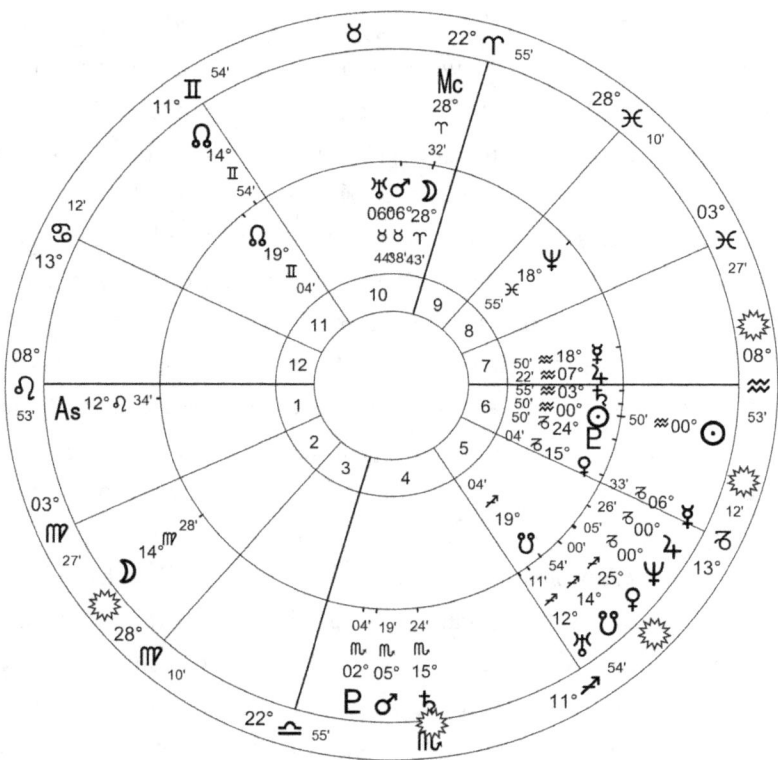

ASPECTS AND ORBS

Orbs

In astrology a variety of approaches to determining the sphere of influence of the planets in a horoscope can be found. There are many approaches, ranging from those that swear by as precise and narrow an orb as possible, to those allowing the Sun and the Moon as much as 15 degrees of orb, those specifying half-orbs for each planet, to some of the few authors, who argue that there are no orbs at all, and that the spheres of planetary influences should be determined in other ways.

Through years of practice, every astrologer has developed a system of his own. As for myself, I generally do not exceed a seven degree orb in solar returns. There are, of course, various factors to be taken into account, from the size of the planets to their function in the horoscope, for instance the Ascendant ruler, even if this be the little Mercury, should still be attributed more importance in a solar return with Gemini or Virgo on the Ascendant, but not in other returns. It may therefore be useful to make a small note of the ruler of the Ascendant in a chart, so as not to lose sight of its more prominent role when studying the whole picture; some astrologers equate the ruler of the Ascendant, also called the 'ruler of the horoscope', with the Sun and the Moon in the chart, while some even consider it the most important factor in the chart.

The principle that the more accurate the aspect, the stronger its impact, is of course inevitable in the solar return technique, so most attention should be paid to the partile aspects, reaching up to 1°00' of arc.

More interesting, however, are aspects with a wider orb, for example six degrees or more. According to many studies, the strength of an aspect should decrease progressively rather than linearly with increasing angle, let alone be the same over the whole orb, or simply cease at a certain point of distance.[8] However, in my own work, I have noticed that in some cases the influence of a planet with an orb greater than 6° appears to be very strong, even if not formed by one of the luminaries or the ruler of the Ascendant, i.e. one of the most prominent elements of the solar horoscope. Since the phenomenon is not consistent, it opens up a whole new field of research. Perhaps just a hint here: in some cases this was the only aspect of the planet in question, or even the only aspect of both planets that were not aspected by the rest of the horoscope, i.e. neither by the solar planets nor by the natal planets, a phenomenon also known in natal astrology as a "duet".

Inconjunction

Conjunctions, oppositions and squares are considered - in that order – to be the strongest aspects, followed by others. In my own work I use the Ptolemaic aspects with the addition of the inconjunction, which is an extremely illustrative aspect in solar returns, denoting not only the need to adapt, but also the level of stress, nowadays quite a "normal" part of our lives. The orb for this aspect is smaller, perhaps up to 3°, but it should nevertheless not be overlooked. This is especially the case if there are several inconjunctions in the solar return, while there is none or only one in the natal chart. This is a

---

8   J. Eshelman reports in this connection that a conjunction with orb of 3°00' is twice as strong as a conjunction with 7°03', the latter being twice as strong as a conjunction with an orb of 8°34›. According to these studies, conjunctions and oppositions should still have some 50% of their power at around 7°, and trines and sextiles somewhere around 5°.

definite indication of a high degree of stress in the current year, and it is up to the client and his character to resist and fight back, or to succumb to pressure. In such a case it is useful to examine specifically the so-called health houses of the horoscope, inconjunction as a rule being an indication of a person's physical or psychological, often mostly psychosomatic, response to pressures, he or she is unable to cope with.

Isolation in the astrological chart

Unaspected planets, i.e. planets that are not linked to other planets through ptolemaic aspects and inconjunctions, are often a rather unpleasant feature of an astrological chart. In a solar return it is therefore a good idea to mark the natal unaspected planet separately, as well as to check its aspects with solar planets. These aspects usually turn out to be particularly strong and significant.

At this point I would add the observation that while an aspect to the Ascendant and/or the Midheaven will not alter the planet's unaspected status, as only a planet can do that, aspects to these two angular points are nevertheless an essential part of the solar return interpretation. Ascendant represents not only a stamp of the current year, but also 'one's self', and the planet it aspects adds expression, support or conflict, depending on the planet and the aspect. The same applies to the Midheaven, i.e. the 'point of realization', the planetary aspects it forms indicating the possibilities, ways or the degree of realization in the current solar return year.

SOLAR RETURNS

Creating configurations

One of the most important pieces of information in a solar return horoscope are the combinations and configurations, created by the solar planets. For example, a natal square between two planets, turned into a T-square by a solar planet, or a Grand Cross completed by a solar return planet from a natal T-square, to name just two of the most obvious possibilities. The natal planets in the new configuration find themselves in different relations and roles than in the natal horoscope, and their strength in the solar chart is now different than before, either increased or diminished. In such situations the original natal configuration must again be taken under scrutiny and examined together with the solar return chart, to assess its role in the new chart.

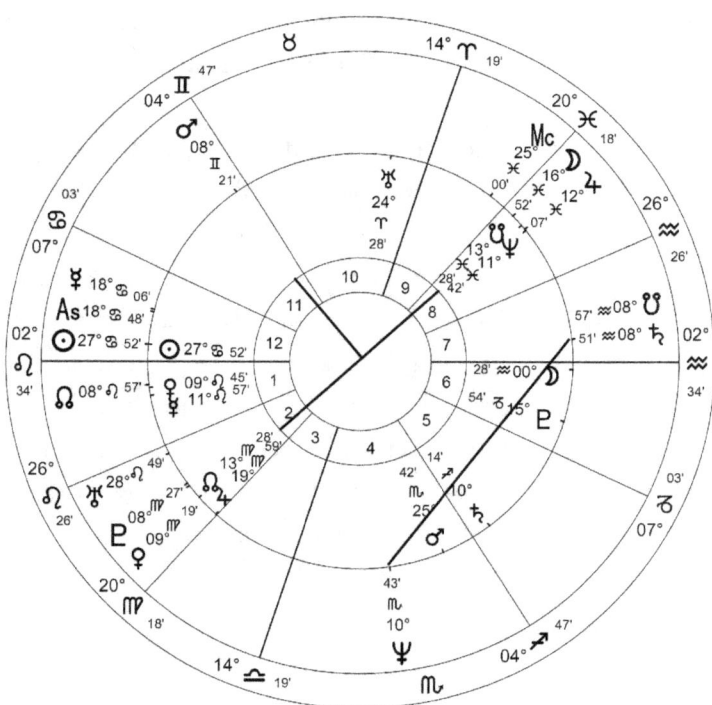

Not just one planet, but several planets can be aspected to a natal configuration, or a specific configuration can be formed in the solar horoscope, which then aspects a natal configuration - the possibilities are probably endless. Let us examine an example of a solar return with such upgraded configurations.

The chart shows a natal configuration of the T-square formed by Pluto/Venus, Jupiter (and near it the Moon) and Mars as the focus of the configuration, plus the square between Neptune and Saturn, together with the South Node. In the current solar return Venus and Mercury have upgraded the square to a T-square, with natal Neptune in the 4th house as the focal point, which is also placed in the 4th house in the natal horoscope, so this is also a repeated position.

The marked T-square received an even bigger boost in the current year with Saturn in Sagittarius opposite the focal point Mars, plus Neptune and a pair of Nodes. Add to this the yearly mark of Leo, and you can imagine how this person's life has been turned almost literally upside down...

The frequency of such cases is evident from the example of 2020, when for eight months the three heavy planets Jupiter, Saturn and Pluto were travelling "side by side" through the sign of Capricorn in a stellium, with a maximum mutual distance of 8 degrees. Fortunately, for the most part, they were not forming very difficult configurations with other planets, apart from the rapid daily motion of the Moon, which turned quite a few squares into T-squares for a short time. However, there was a case of five planets in a T-square on 1 August that year, and even six planets in a T-square on 30 April! Obviously two thirds of the population had this difficult stellium in their solar charts! We all have birthday once in a year and eight months make two thirds of a year. In addition, a large number of people have planets, apart from the Sun, or aspects to planets, somewhere in cardinal signs, which were activated in one way or another by this difficult stellium. No wonder then that the stories of that year were more tragic than usually.

Particularly prominent are all aspects – and even more so configurations – relating to unaspected planets in natal chart. Every unaspected planet very much "yearns for" an aspect, even if unfavourable or fatal, as long as it is not alone… It is therefore useful to mark an unaspected planet in a solar return chart and to check all the aspects it makes with solar planets.

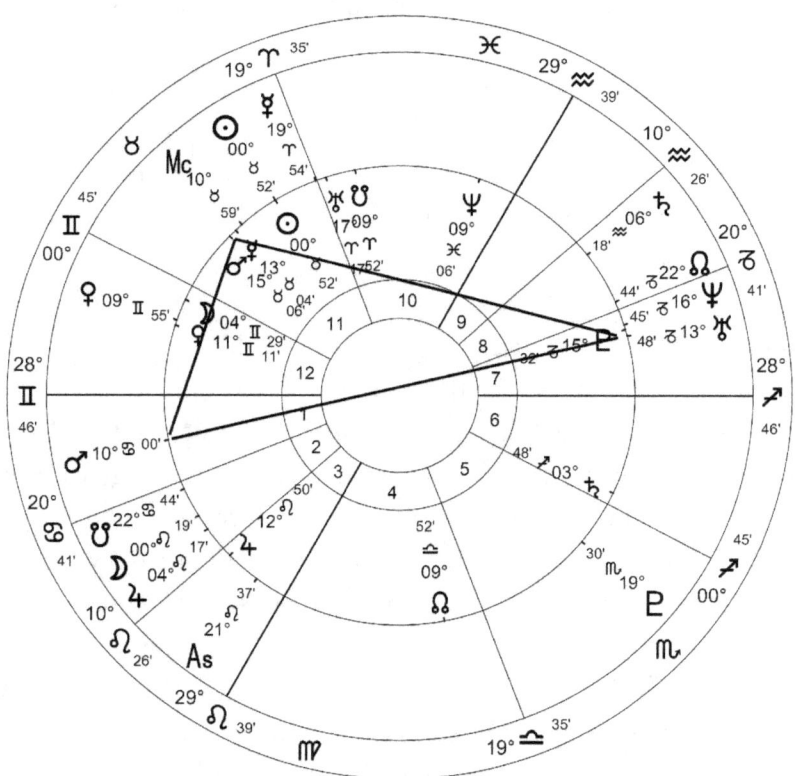

There is another phenomenon worth mentioning with regard to these new possibilities. Namely, a planet that is found at an angle of 60 or 120 degrees to an unfavourable position or combination can safely be called "the way out", offering the possibility of resolving a difficult situation with its help - whatever its position may mean or entail in a particular case. A sextile and/or a trine (ideally both) aspecting a T-square or even a Grand cross are two most welcome elements in a horoscope... The enclosed picture of a case shows the positions of solar Jupiter and Saturn 'offering' a 'way out' of a very unfavourable natal opposition between the Moon and three planets, the Moon having no favourable aspects in the natal horoscope.

It is therefore quite possible that such a newly acquired combination or configuration with a 'way out' will not be experienced so intensely; what is more, it may even be the case that this very 'way out' will affect the energy of the whole configuration and make things turn out quite differently from what we had anticipated or feared.

COMBINATIONS OF HOUSES IN A SOLAR RETURN

Sometimes, the astrological house corresponding to a particular area of life is not sufficiently indicative of activity and other relevant houses have to be considered. Such is the case of the already mentioned relocation, for which combination of at least three houses in the horoscope has to be examined, namely the third (change), the fourth (home, property) and the sixth (order, lifestyle). Business matters, divorce, property sale, etc. are all associated with their own particular house combinations.

However, planets, placed in these houses, are not a sufficient indicator. With ten planets spread over twelve houses this way or another, there is a fair chance statistically that at least three will be found in the three houses associated with relocation. However, the potential for relocation will not be at its strongest, unless they are interconnected or intertwined. It is therefore most favourable if the ruler of the third house is placed in the fourth or sixth, the ruler of the fourth in the third or sixth, and the ruler of the sixth in the third or fourth. [9] It does not help much, though, if the ruler of the same house is in one of these houses, for example the ruler of the sixth house being in the sixth, as the latter is also related to work, hygiene, health or small animals. There is no connection. In addition to these three rulers, the ruler of the first house or annual Ascendant can also be helpful, itself highlighting one of the houses, needed for the combination.

A similar case applies to the health indicators, except that the combination here is 6 - 8 - 12, necessarily including the ruler of the Ascendant, i.e. the first house, which, as a rule, shows one's body and its health. Even the most "ominous" and "at first sight alarming" eighth house will not necessarily bring health problems, surgery or death, unless clearly related to the rulers of the sixth, twelfth or first house. It will rather be a kind of an ultimate warning, a search for truth, depth, the occult, a break with mother (if containing a badly aspected Moon), but the issue will not be health or surgery. The same applies to the sixth house. Only combined rulers will point to increased health activity.

A special mention goes to the twelfth house. Heavily occupied, it is always a cause of anxiety. Before informing client about depres-

---

9   This does not refer only to the primary ruler of the house, but also to co-rulers, all the more so, if activity takes place later and not exactly at the beginning of the solar year. More importantly, the same houses may also be occupied by their natal rulers, which in turn will contribute to the combination of houses, needed to fulfil the potential.

sion, introversion, hospitalization, or anything else pertaining to this particular house, it is necessary to assess, whether this is - mostly or entirely - an active or a passive connotation. This refers to the sign on the cusp (active-passive in the order of Aries-Taurus, etc.), the placement of its ruler (even if in the sixth house!), and the active/passive quality of the planets in the house. For there is a very big difference between staying in a hospital as a patient, or working there as a doctor, nurse, administrator, etc. A strongly occupied 12th house may also be an indication of energy and time investment into personal growth, courses, therapies, etc.

An interesting case is when both the 4th and 11th houses are strongly occupied or highlighted in the same yearly horoscope or in two consecutive solar returns. These two houses have nothing whatsoever in common, being practically mutually exclusive in at least one quality - the fourth house often refers to the retrospect, for example, clearing up (relationships or affairs) at home, or clearing out old stuff (mentally or materially), while the 11th house is characterized by focus on future, perspective, vision, ideas, light at the end of the tunnel... Often these two principles in the horoscope point out the way forward after things that needed clearing up had been resolved. Sometimes this even occurs in the course of the same solar return, with the fourth house then being busier at the beginning, and the eleventh house later in the year...

There are of course many more such combinations and intertwinements, and it is difficult to list all of them without forgetting many practically useful examples. To this end, Appendix 1 shows some typical combinations, although just as cues, not as real recipes.

Besides, every practitioner – like in cooking - finds his or her own tricks and shortcuts in time or improves the existing ones.

ONCE AGAIN "SOMETIME IN THE YEAR?"

Several prominent astrological authors, who had studied solar returns extensively, including Raymond Merriman, Mary Fortier Shea and Marc Penfield, suggest that "a solar return horoscope should be valid for more than one year", for example for three months before the return and for three or even six months after the end of the respective year.

There are two nuances of expression here, namely that *"the solar return lasts for more than one year"* and that *"the solar return effect lasts for more than one year"*, which are not quite the same; the authors are not in complete agreement here. Both claims can be regarded as elaborations of the generally accepted theory that *"the planetary effect in a solar return should occur once within a year, even if on the very last day of the solar return"*, regardless of the activity taking place at the very beginning, end or sometime in between.

As to the second claim, we can agree, as it coincides with our own findings, which will be discussed further in the book. There are events in our lives that are too imprinted on our memories to forget them, having repercussions stretching over a long period of time, or "having a long beard" in popular parlance. A particularly powerful or significant event (or realization), occurring right at the end of a solar return, is likely to have an impact on our perceptions and actions even **after the** birthday, i.e. during the period, already covered by the next solar return. The duration of influence, however, depends mostly on individual.

The first claim raises the question of how can a yearly horoscope be called "yearly" if valid for a year and a half. However, the more important question here is, that already in the very next step we encounter a potentially unsolvable problem, since we will then **also** be studying **the next** solar return as a period of 18 months, three of which taking place in the current solar return year. The same applies to all the future solar returns. In simpler terms, if this theory were true, some three months of our life could be reflected in two successive yearly horoscopes, which is a paradox, because the first **and** the last three months of each solar return could then always be visible in two successive charts. What happens if the two charts are contradictory? Or more specifically: if we take the example of a troublesome marriage, the interpretation being valid for over a year, would it then be logical to observe serious problems in two successive solar returns? It probably makes sense to reduce the solar return and its effects to the twelve months it covers and to stop bothering.

A mark right at the end of the house in the solar return

However, there is some accuracy in this theory, as evident from practical work with solar returns. Here's the thing: the position of a planet at the very end of a particular solar return house can even show us, what had happened just before the respective solar year! In other words, that event should have been shown at the very end of the previous solar year, not in this one!

At the first sight this is an illogical statement, but if we consider the mentioned principle, along with the often mentioned and crucial aspect of human **perception** of the horoscope, it is possible that an event is so important as to denote a longer period, say a few weeks.

And if we have birthday immediately thereafter, or such a significant event occurs shortly before it, such a perception may accompany us as some important circumstance or feeling into a new personal year and thus into a new solar return.

RECTIFICATION

Whenever the time of birth is unknown, inaccurate or unreliable, astrologers can use one of the many methods of determining the accuracy of birth time, known as rectification. There are many methods, a major part of them based on collecting as reliable as possible data about important or key events in a person's life, along with determining which astrological shifts, contacts and activations correspond most closely to these events.

Searching for the time of birth in the absence of any information or in case of it being too general, e.g. "morning", is time-consuming to the extent that the efficiency of the rectification is questionable; however, when it is merely a correction of a not very wide time range, there is more chance of success.

If solar returns are practised according to the method of intra-annual timing, presented in this book, they turn out to be a very useful tool also in the case of birth time precision.

Let us examine two examples.

Was it really at „12.00"?

The "12.00" birth time in itself is enough to cast doubt on its accuracy. It is recorded in the birth certificate, but since miss Tájana, whose horoscope we are discussing, comes from a background where, at the time of her birth, this information was not yet consistently recorded, one may question its accuracy. An astrologer can undertake this check by rectification, although it can also be done with the help of solar returns, as well as without further calculations.

On 17 April 2013 my good friend Josip, Tajana's brother, passed away suddenly. He was one of those unforgettable people, who truly remains unforgotten.

His story could be presented at length (and deserves to be), but we will only deal with the essentials of his case. Fifteen years ago he was involved in a fatal accident as a passenger in a head-on collision, flying forward from the back seat and suffering multiple fractures of the base of his skull, as well as a loss of his left eye. He was told that practically every remaining day of his life was a gift to him, and that is also how he lived. Later, he began to suffer from attacks of a type of epilepsy unselfishly with increasing frequency. However, Josip continued to live as well as he could, helping others, being positive to the end. Then it 'hit' him one morning and he collapsed in the bathroom. The extremely quick medical help came too late. The cause of death was not an epileptic seizure, but a cardiac arrest.

If we agree with the theory that our own death cannot be seen and should neither be sought in our own horoscope, it is nevertheless often reflected in the charts of those close to us, in this case Tajana's, who was deeply shaken by the loss. In her horoscope the brother is associated with the fifth astrological house and not with the third, which is assigned to the elder sister. The ruler of this house is Saturn, placed in the eighth house and close to the Moon's Node, also squaring the Moon.

If the given birth time is to be accurate, at least some of the elements of the solar chart that can be linked to the brother and his death, should accurately reflect the time, indicating his death.

Tajana was born on 16 May, and 17 April is 29 days before the birthday, meaning that we will be looking for points in the astrological houses of the annual horoscope for her year 2012/2013 at a distance of approximately 11/12.

The Ascendant of this solar return is in Aries, and Mars, its ruler, is placed in the fifth house, attributed to the younger brother. Mars is placed at the very end of this house, spanning 35°32, which means that a month in this case measures a little less than 3° (actual calculation: 35°32 = 2132'; divide by the number of months in the year: 2132' : 12 = 177.67' for the month. Conversion into degrees gives the result 2°57'40"). The end of the fifth house is at 12°37 Virgo, and the position of Mars is 9°03. The difference is 3°34', showing considerable precision in proportion within the solar return horoscope.

In this horoscope, the Sun, not only as the 'main element of the solar return horoscope', but above all as the ruler of the examined fifth house, is placed at the very end of the first house, i.e. angular and thus even more prominent. Here again it is found at the very end of the house, ending at 27°45 Taurus, with the Sun's position is 25°37 of that sign. The difference is 2°08, with the house itself measuring 33°45, almost the same as the fifth house calculated earlier. The result is again "about a month before your birthday".

A third element, related to death of the younger brother in this chart, is the position of natal Saturn, the ruler of the fifth house (younger brother) in the natal horoscope. It is at the very end of the 12th house, measuring 41°23, the difference between the position of Saturn and the Ascendant totalling 3°39. 1/12 of this house is calculated as 3°26'55".

All the three results are relatively close together, which illustrates sufficiently the usefulness of proportional timing in solar return horoscope. However, there are two more planets in this chart, similarly placed at the very ends of the houses, which only further highlights the prominence and importance of "about a month before your birthday" - the position of Pluto in the ninth house, which in the system of derived houses rules the brother's fourth house, i.e. the house of endings, and natal Neptune in the seventh house, whereby Neptune in her natal horoscope is found in the third house (siblings) and also in Scorpio. In the natal horoscope Neptune forms a T-square with Sun and Jupiter, this T-square being highlighted in the current solar return by Jupiter, placed right next to the Sun...

So, if we accept and master the proportional model of timing events in a solar horoscope, we can claim with a relatively high degree of certainty that this lady's birth time of "12.00" is correct, even if we had been sceptical about it.

With just a few attempts to apply a different birth time - something we can afford in the age of computers - we can see, how the positions, reflecting a strongly indicated event, would shift, practice proving that a five-minute shift in birth time shifts indications in a solar return horoscope by two to four weeks, i.e. by a further 1/12th of the extent of the house, containing the planet of observation.

Ten minutes

Although there are cases in practice, where a planet or a house cusp changes sign even within a minute of difference, in most cases with only a few minutes difference in birth time, the birth horoscope changes little if at all. Often we only find out that something does not match in the prediction process, where even small changes in horoscope can significantly shift the predicted dates, for example in secondary progressions. However, to find this out, we need as accurate times/dates of significant events as possible.

The horoscope of a man, named Ivan for the purpose of our illustration, is a good example of solar return having indicated the need to correct the birth time. His birthday is recorded as "23.15", in his mother's memory "a little after 11 p.m.", which is quite accurate.

Review of events from the recent past was not difficult until we reached the time around his birthday in 2022. In fact, quite a few planets were perfectly aligned with the house cusps, suggesting considerable dynamics on the birthday itself or in the days around it. Let us have a look at the charts:

Natal Pluto is exactly on the MC, Uranus is exactly on the cusp of the 6th house (both indicating job and work), and the Sun is placed just after the Descendant, a few days after his birthday. But to my amazement, Ivan was not only aware of the events, but also of their timing, which was a good month after his birthday.

This realization in itself forced me to make a correction; with some calculations and simulations I came up with a birth time of "23.05", which seemed most accurate. In this way I was able to time the three positions more precisely to 'five weeks after the birthday'. According to the calculation of the proportions within houses, which we mastered in the previous chapter, Pluto (3°18 from the house

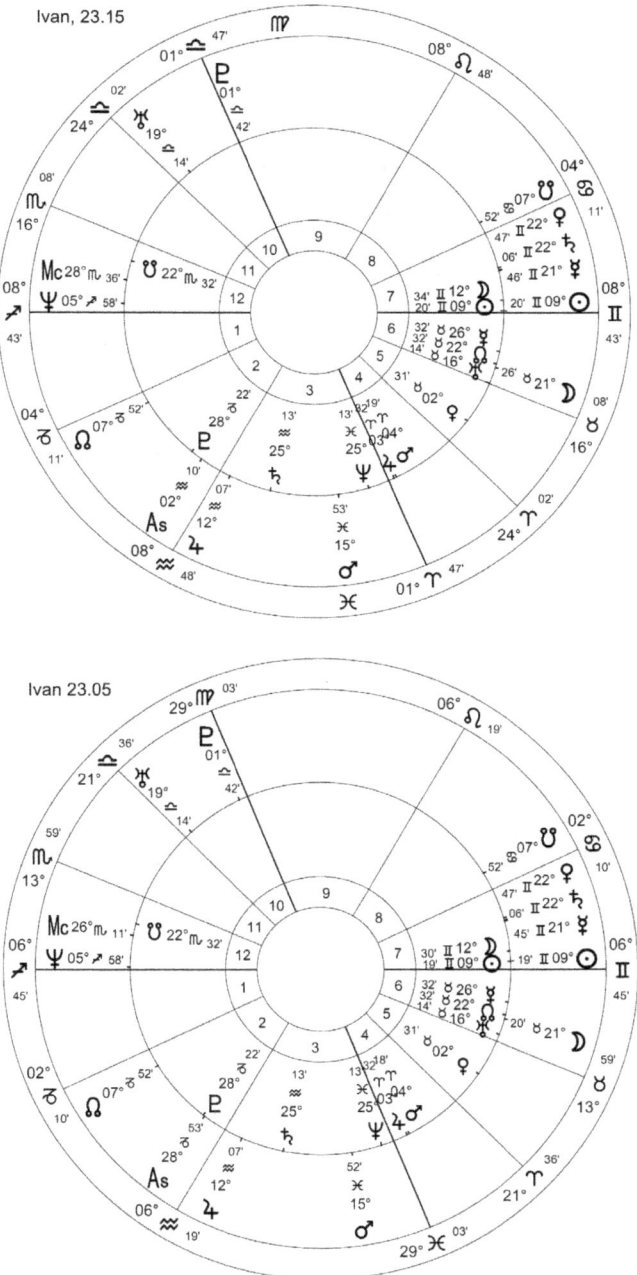

cusp) is placed some six weeks after the birthday, natal Jupiter in the third house (5°26) about five weeks, Uranus in the fifth house (2°01) about one month, and the Sun in the seventh house (2°17) also a little over a month after the birthday. All four at almost the same time! The rectified birth time then proved to be reliable in the verification phase, having timed some key events in life much more accurately than the original '23.15' birth time.

And what happened? Natal Pluto brought an inspection, a demand to react quickly, lots of short trips and paperwork (Jupiter in 3rd), a halted or severely disrupted work process, resulting in the immediate dismissal of one employee (Uranus in 6th house), and very shortly thereafter a quarrel with a business partner (Sun in 7th house, already squaring Mars in natal chart - potential for disputes). Mars here is placed in the fourth house, which brought about conflict with partner, not related to the inspection, but to the accumulated things they had not cleared up in the past.

ON PREDICTING

Regarding the accuracy of calculating, let us have a look at some thoughts by the Croatian astrologer Slaven Slobodnjak, taken from his treatise on mathematical precision in astrology and published with permission.

"In today's astrology, the ideal is to find the most accurate aspects possible in prediction - we are delighted to find an event correlating exactly to a transit. But how often does this happen? Does the accuracy of calculation really increase the understanding of astrology and human beings?"

The author goes on to describe the tendency of searching for something precise, just to confirm mathematically and as accurately as possible an event we have come across: *"And somewhere we will finally find something clicking, and we will say 'Oh, that's it...'* Well, if we are a little critical, we will have to admit that precise aspects rarely ever produce events – that is why we use orbs, thereby only adding to the number of possible factors."

And finally: "*Sometimes this quest for precision seems to generate situations, where we 'can't see the forest for the trees', failing to distinguish the essential from the non-essential...*"

VII

LUNAR AND PLANETARY RETURNS

The Sun's return to the exact same position of the birth horoscope is just one of the cycles that can be taken into account in astrological work. With everything in nature being cyclical, all the planets travelling in their orbits around the Sun, and astrology being but one of the most illustrative demonstrations of this principle, it is logical for the planets to return sooner or later to their natal positions. Hence it follows that astrology also deals with lunar return, and then in planetary order Mercury, Venus, Mars, and so on up to Pluto's return. With each of these bodies a point in time can be found, where its position equals its starting, i.e. natal position.

The practical applicability of these calculations and return charts is, however, questionable. The trans-saturnian planetary returns seem somewhat unrealistic due to their orbital time; part of the population may experience, for example, a Uranus return, however, the significance of a "horoscope for the next 84 years" (which is what a Uranus return chart is) is a purely rhetorical question. The impact of Saturn's return on a person's life is not in the least controversial, since the ages of 29½ and 59 are, in most cases, key points of personal development and growth, as well as karmic fulfilment. But will a horoscope for the full 29½ years of our lives provide sufficiently clear clues for an astrological interpretation of the qualities of this time-spanning period? Even a Jupiter return, i.e. the horoscope for almost exactly twelve years ahead, raises a similar dilemma, along with the question of advisability of casting a "horoscope for the next twelve years".

The horoscopes of personal planetary returns are infinitely more relevant for client's perception, since they all expire in a realistically manageable time, somewhere up to two years; only Mars, with retrograde period involved, may take a few months longer. However, none of the personal planetary returns takes too long for a person to remember the key events of its respective period. It is the retrograde motion, however, which, in the case of personal planets, influences the length of the respective return periods. But even if we simplified the matter and make Mars' orbital period simply "two years", it would in practice sometimes be difficult to interpret the potential of this pair of years, along with the potential of the next pair - say, "what 'pair of years' are we in now"?

Unlike these, the lunar and solar return durations are predictable, of almost exactly the same length all the time, and therefore most frequently used in astrology. Many a practising astrologer does not use planetary returns in his work, preferring to use returns of the two luminaries.

Let us therefore take a closer look at lunar returns.

LUNAR RETURN HOROSCOPE

The Moon's orbital period is 27.32 days, also called a 'sidereal month', meaning that it makes more than 13 orbits in a year, i.e. returns to the exact natal position of the Moon (a more accurate calculation gives 13.37, calculated over 365.24 days as an average long year). A lunar horoscope is a representation of activity and perception within one lunar month in our life, from the viewpoint of emotions, feelings and everyday matters. Even if the Moon is extremely prominent in a person's horoscope, a lunar horoscope is not likely to indicate a radical change or upheaval in a person's life. It could rather be said that this horoscope reflects our current mood, reactions and the surrounding atmosphere. Rather than a picture of a crisis or a pivotal event, the Lunar Return horoscope provides information on our perception and emotional response to the crisis.

But a word of caution here - if it is an event, indicated in the current solar year (and therefore in the solar return horoscope) or even in the natal chart, a lunar return chart will give us an even more accurate indication of the respective event and the surrounding circumstances, as well as the degree of emotional involvement and reaction to the event itself. The lunar return in such cases proves to be an excellent timing tool, confirming the unfolding of anticipated or predicted event in the current month, as well as showing peak intensity within a long lasting process.

Despite high-speed computers, capable of calculating a chart in a fraction of a minute, it is not advisable to cast all the 13 lunar horoscopes within a year to determine timing. This technique, however, is best used in combination with solar returns, progressions, transits and other predictive methods we normally use. Practice shows that by studying and comparing a sequence of lunar returns, it is possible to detect the appropriate time for the fulfilment of a potential, and above all to determine the lunar return, in which something predicted for an approximate time by some other technique, is most likely to unfold.

The word 'approximately' is of course not used by accident – a Saturn transit, for example, can take quite long, especially if stationary and changing direction, when it won't move significantly for some time. The same applies to progressed charts, where with planets staying within orbit of a planet or point for a considerable period of time, it is often difficult to determine, timewise, the fulfilment of the potential, implied by a position. We therefore need a method to narrow down the timing of the potential fulfilment to such an extent that predictions will be more accurate. Lunar return is neither the only nor the best tool for this purpose, however, it can be successfully used in terms of a more precise timing of an activity, indicated in prediction.

Some useful tips

A lunar return horoscope has the same set of elements as a solar return horoscope, but their interpretation differs. Let us therefore mention the most important points.

(Note: In the survey of these indicators, I use the term "month" because it is simple and convenient, however, readers should at all times be aware that lunar return does not refer to a month of days we are used to, but to the time between two equal positions of the Moon, or the time of one revolution of the Moon's path around the Earth, with the additional complication that lunar months start anytime within month. This is because, over the year, the return dates within months move slowly backwards, like the positions of full moon, if recorded on a year-round calendar.)

The Moon's position in each lunar return will naturally be in the same sign, but its house placement will vary. While solar returns follow a certain - albeit highly complex - sequence, which is even predictable to a certain extent, this is not the case with lunar returns. Even a brief examination of the successive horoscopes of the lunar returns shows that the Moon's house position is completely accidental. Even more crucial are its aspects, varying from month to month and generating the distinctive dynamics of the lunar returns.

Another important piece of information not to be overlooked is the Moon's house position, the house with Cancer on the cusp, as well as the fourth house that is "naturally" ruled by the Moon. In a sense, these three areas will provide in the respective month, perhaps through emotional expression, interrelatedness, personal energy, some general picture of the emotional state and activity of the month.

The two houses containing the natal Ascendant and Midheaven are the two most outstanding areas in terms of activity and personal progress in a current lunar return chart.

If it is possible to "assign" a quality to a month, in other words, a summary information, it is the lunar return ascendant that sets the tone of the month in terms of emotional perception. Of course, the most important months in this regard are the ones, where lunar and natal Ascendant fall in the same sign, as this triggers certain potentials, acquired at birth. In such a month some of the birth potentials may well be manifested or expressed. This is especially true if natal and lunar Ascendants are less than 5° apart. Generally speaking, the sphere coinciding with the Moon's placement in such a chart, will be coloured by our emotions and inclinations.

Simultaneous use of solar and lunar return

A lunar return describes influence of the planets (planetary energies) on our emotional, inner life in the course of one lunar revolution. Such a horoscope clearly indicates so events (as we experience and perceive them personally and emotionally) as our reactions.

The Moon is an important indicator of personal fluctuations and emotional perception of everyday challenges, which makes lunar returns more useful for interpreting short periods than for exploring a trend or evolution of a situation, as was the case with solar returns. The two approaches can be very complementary, bearing in mind that the two charts cannot be interpreted in exactly the same way. Each of the horoscopes reveals its own perspective, but together they give a surprisingly accurate picture of the perception of the observed period. In this respect, a solar return can be used as a basis to determine the part of the year, relating to the activity we are looking for, following or planning.

With this purpose in mind we look for the lunar return that is most closely linked to the potentials of the respective solar return. All the peculiarities pertaining to other predictive techniques, such as aspected positions (between the planets of the solar and lunar returns), repeated positions (and especially repeated positions from the natal horoscope!), the same Ascendant sign and so on, will also relate an important information in the case of a lunar return horoscope. In a survey of a series of lunar returns, the "one" most closely corresponding to the solar return potential will almost certainly stand out, and will be easy to notice and pick out.

It is even easier to work with a computer program that places the lunar return around the solar return, helping us quickly find the month that best matches the solar return potentials.

However, in astrological practice, where there are supposed to be as many approaches as astrologers, the opposite approach can also be found, namely the use of solar return as a kind of connection or explanation of a series of thirteen lunar returns in the same year. However, such work is very time-consuming, and in studying thirteen astrological charts over the course of a year one can lose track of the real picture.

SOME EXAMPLES OF USE

Birth of first child

Petra is a tourist worker and a former model, who gave birth to her first child, a son, on 17 October 2011. According to her birth date of 2 May, 17 October is just before the one-half year horoscope, so in a **solar** return horoscope we would expect to see indications of motherhood at that point, which for most women is not 'just' the birth of a child, but more - a change of status from a girl to a mother. Moreover, this pregnancy clearly meant the end of the modelling career. This is indicated by Saturn, placed in the 10th house, exactly on the degree where, by calculation, we would expect a change on 17 October - responsibility, seriousness, career stagnation. In the same places within the houses, i.e. a little before half a year, the natal Moon (in the 6th house), Neptune, ruler of the intercepted 3rd house (in the 3rd house), and the natal Ascendant in the 8th house are found, denoting a powerful change of higher significance.

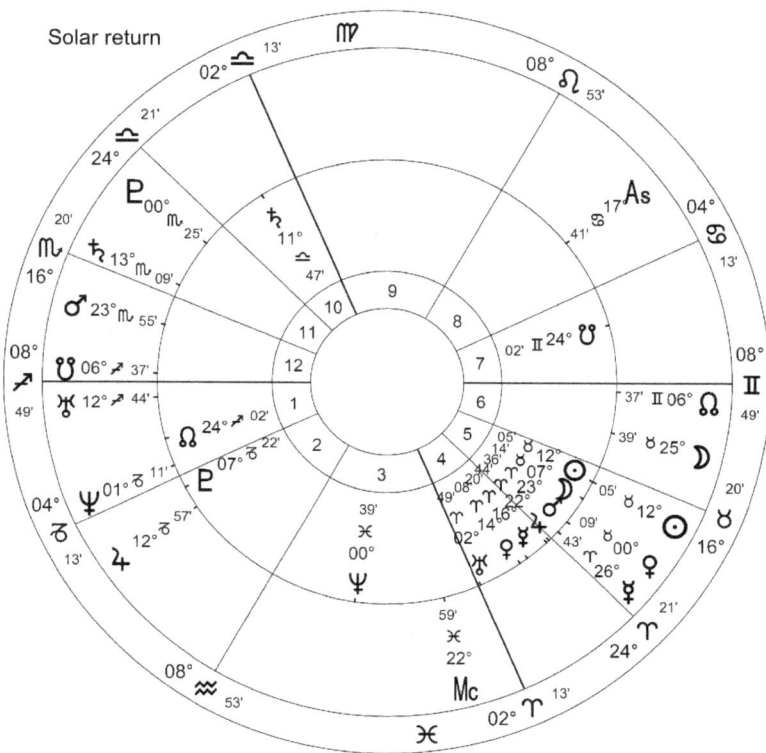

We would expect such an important event to be reflected in the horoscope of **the lunar** return. The Ascendant of this lunar horoscope is Libra, and its ruler Venus, together with Mercury (associated with the concept of children it is also the leading planet) is placed in the 1st house – the meaning of change for the person. Both Moons, as the key element of this chart, are found where the Ascendant was in the solar return – in the 8th house. Since the client gave birth two days after the onset of this chart, if the timing we have observed in solar returns is to work in lunar returns, some important indications should be found at the very beginnings of the astrological houses. And there they are: the Moon at the beginning of the 8th house, natal Mars at the beginning of the 2nd (the house of money, but also of values and worth) and Neptune at the beginning of the 5th house, the house of children.

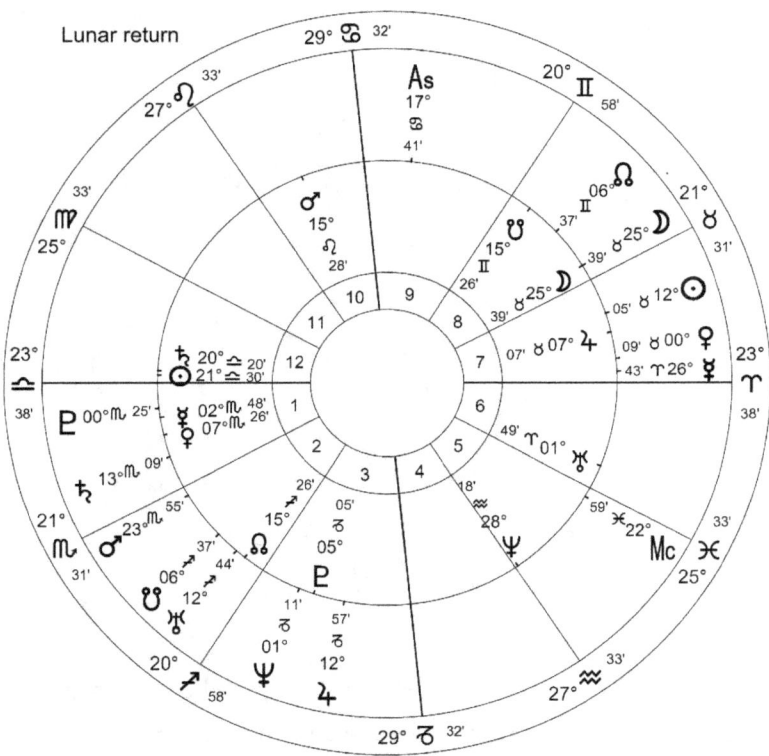

But are these the indicators one would expect at such an important life event as birth of a first child? They may not be the most expected ones, although they certainly have a meaning, given the timing in the horoscope. At least the Moon (8th house signifies a major turning point) and Neptune (a wanted child, but nevertheless a radical change of lifestyle) are easy to understand.

The lesson to be drawn from this is that a lunar return can only be an auxiliary means of determining the time (or rather: narrowing the time) of an event, as well as its indications. The planetary positions in the lunar return horoscope are more likely to indicate emotional response to an event, rather than its quality.

If we imagine everyday events from different perspectives, an event may excite us in real time to such an extent as to seem overwhelming and crucial, yet only after a few days it will no longer mean anything special, depending of course on the individual's perception. As an illustration, let us imagine a typical Taurus, who had misplaced his well-stocked wallet. He would be more depressed by the potential loss than any other sign, and would devote all his attention and energy to finding the money that simply must not be missing! A search could take days, especially if the wallet was last seen in an unusual place. Quite a painful matter and a few days of severe anxiety. Then suddenly the Taurus finds the wallet with all the money still there - a greatest relief! He won't think about the money again for a day, and in a few weeks he won't even remember it, if asked about the outcome of his intense search. There was no problem below the line – the money is there and life goes on. But during those few days when the wallet had been missing, his neighbour might have fallen ill, his child might have hurt his knee, his aunt might have come to a visit, or something else, but our Taurus would only have worried about the loss of the money. This is just a theoretical case, and not all Taurus-born people are so obsessed with money, but it does illustrate well the difference between our emotional perception and reality.

Example of a series of lunar returns

Lunar returns therefore show the tone of our perception of each lunar month, helping us to discern some of the main highlights of the month, but less easily the events as such. Looking at an event from the distance of time, we are only left with the main landmarks and outlines of the respective time. If we do not have a record of what had happened, preferably with time notes, memories will slowly start blending together, and we will lose track of events and their timing.

But if there is a whole series of important events that we are not likely to forget, the lunar return horoscopes can also help us see what it was all about. Let us examine the case of a good acquaintance of mine, namely the six months of his life in the summer and autumn of 1989 and the transition to 1990.

By July 1989, his marriage was already on shaky ground, and some emotional upheavals of those days led him to a deep disappointment, even to doubts about the way forward. In the LR (lunar return horoscope) for the month between 29 June and 26 July of the respective year, depression and introversion are visible (Moon in the 12th house, opposition to Pluto in 06, and a partile square to the MC/IC axis = disruption of order, structure), and the South Node is in the 4th house together with Pluto. From here it is not far to the definition of "my wife is tearing down my home", and that is exactly how my acquaintance described his feelings to me at the time.

The following month (26 July - 22 August) was marked by a dilemma about children - information came that the chances of a possible second child were fading due to health reasons, and that a decision for a new family member would have to be made practically 'right now'. The Moon of this return is in the 9th house (this would otherwise be a third pregnancy) and in square with Mercury (ruler of children) in the 11th house, together with the Sun - the man was more focused on future than on the marriage itself. A square between planets does not bring ease, and things were indeed not easy. Uranus is also present - "we're not going to save the marriage by getting pregnant..." The ruler of the 5th house (children) Saturn is wedged between Uranus (opportunity, suddenness, urgency) and Neptune (dilemmas) in the 4th house (home, security; procrastination). In the end, despite many talks, the pregnancy failed.

The period between 22 August and 18 September is most characterized by a 10-day training course, organized by the employer and planned six months in advance. Having taken place far away, the acquaintance also used it to retreat and sort out his thoughts and plans, so one would expect a strongly highlighted 9th house (foreign countries, relief) in the horoscope. True indeed: although we do not find the Moon there, the natal Sun (and nearby elements of the birth horoscope), Mars, Mercury and the South Node (the "easy way out", avoidance) are in the 9th house, with Ascendant in Sagittarius - all pointing to the crucial importance of a stay abroad.

Return home – in the first two days of the new lunar horoscope – was followed by some long and mostly agonizing discussions about saving the marriage, which failed, and at the end of September there was a meeting with a lawyer, who was arranging the divorce. The horoscope for the period from 18 Sept. to 16 Oct. thus brings Gemini Ascendant (change), as well as the Uranus/Saturn duo in the 7th house (marriage). The Moon is positioned on the cusp of the 11th and 12th houses - on the one hand the future opening up, and on the other the feeling of defeat, failure.

The divorce eventually took place in mid-November. The lunar return on 12 Nov. should have contained some key information at the very beginning of a relevant house - and in this horoscope Pluto is indeed found (demolition) at the very beginning of the 4th house (home)!

Meeting the stars

My own case from almost the same time - the arrival of astrology in my life – is certainly an illustrative example of the solar and lunar return.

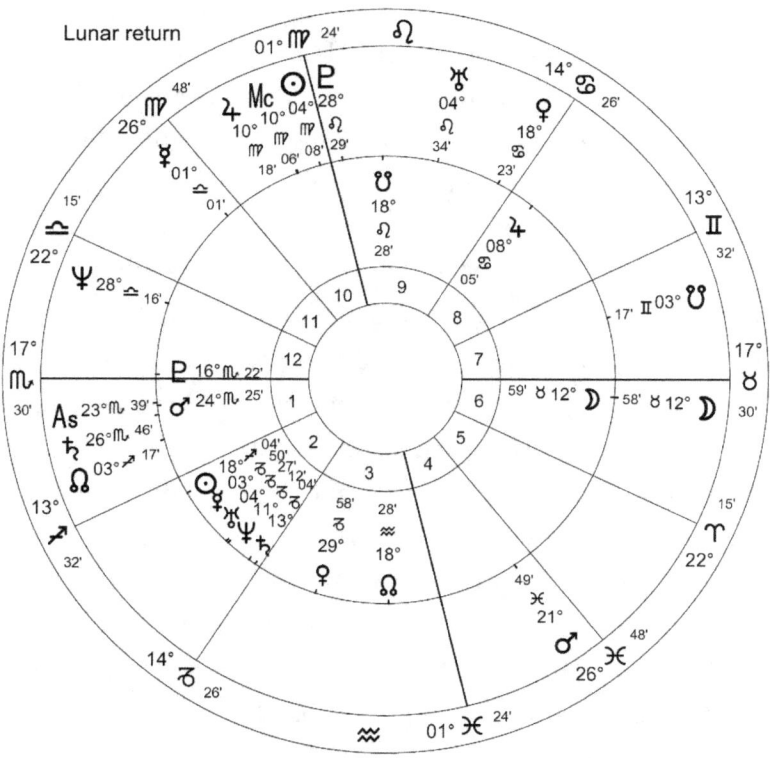

The date is not disputable - the night of 30-31 December of the year, which will therefore be recorded in the lunar return between 10 December 1989 and 6 January 1990. This horoscope is really strong - the Ascendant very close to the natal Ascendant (a crucial period!), with the natal Sun and Pluto on the MC itself (career), Pluto conjunct the Ascendant, Mars as the leading planet in the 1st house (endeavour, energy, enterprise) at only 11 minutes away from the natal Ascendant (!) (so both rulers of the Ascendant closely adjacent to the Ascendant), plus a full 2nd house (value, solidity). An announcement of something new, something big, although in those days of searching for answers to persistently recurring questions, I didn't perceive it that way yet...

The lunar return that follows has a full 9th house, suggesting a search for further information and buying what astrological literature could be found, but everything else is a more familiar story...

SOLAR AND LUNAR RETURN HAND IN HAND

Many events can be observed and explained from different angles, i.e. using different astrological techniques. Thus, one astrologer may first check planetary transits, another secondary progressions, yet another solar returns, profections, and so on. However, the same event can often be accurately described and interpreted by both, solar and lunar returns. Let us examine an example.

On 6 July 2016 Mrs T. suffered from a kind of collapse, focused on her heart, anxiety, fears, thoughts of death and so on. It should be noted here that eleven months before she had already experienced the first such collapse in a series, which is why one can expect each subsequent heart problem to be worse, as well as visible in her horoscope.

In the **solar return** horoscope natal Saturn, general ruler of health, is found in the 8th house; Saturn is also the ruler of the 2nd and 3rd houses, thus ruling the Moon, which is there as the ruler of the 8th house, where (and also in Cancer) Saturn is placed (mutual reception). Calculating the exact position within the solar year does not provide a satisfying result, Saturn's position falling around the beginning of August, almost a month after the event. Such a powerful event is obviously indicated by some other element in this horoscope.

Obviously, at that time Mars is most highlighted of all the planets. Its position in the 10th house is supposed to signify something related to job or career, however, the 10th house is also about status and self-perception through achievements. At the same time, Mars in this horoscope rules the Ascendant, the 12th house, the Sun and Mercury (ruler of the 8th house), is placed in an angular house, the 10th, and is also in detriment, putting the whole activity in a less pleasant perspective. A brief calculation, we are already familiar with, is as follows:

The 10th house of the solar return between the two 19th of November lies between 19°02 Virgo and 12°42 Libra, so the range of this house in the horoscope is 23°40'. One month of this house is 1°58'30", and the month being observed lies between 2°49 and 4°48 Libra. Mars' position is at 3°49 Libra, almost exactly halfway through the month, and exactly one day later, on 5 July. The collapse occurred on 6 July!

LUNAR AND PLANETARY RETURNS

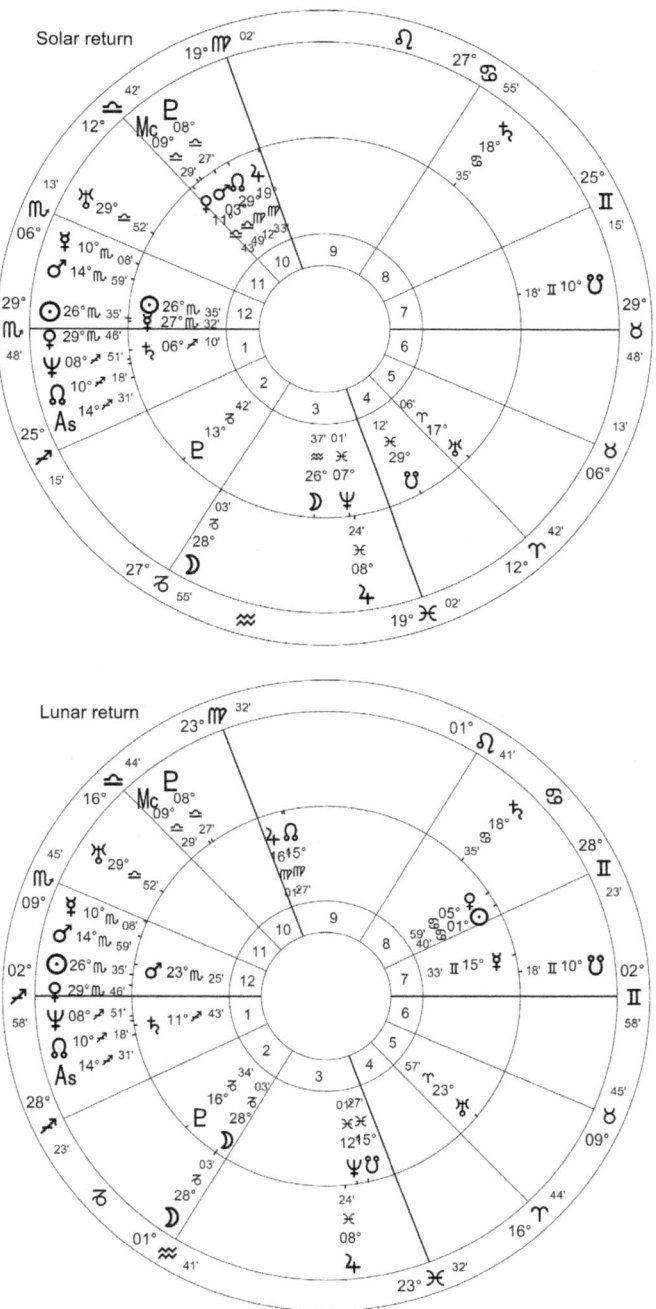

The same activity is also reflected in the horoscope of **the lunar return**. For a collapse, "the moment of a total crash ", Pluto must be involved. Especially if in or conjunct the twelfth house. In the lunar return of the month of the collapse, Pluto is not exactly in the twelfth house, but in the second (value, stability), and is therefore the ruler of the twelfth, which in turn contains Mars, both of them ruling the Sun.

But this chart is even more illustrative, with seven of the twelve houses having a planet somewhere halfway through. And if the timing in a lunar return "works" like in a solar return, we logically expect this "half-house", and therefore "half-month", to coincide with the date of the collapse. So many indicators/triggers at the same time must surely have a special significance!

This chart covers the period between 22 June and 20 July, with 6 July being exactly in the middle of this range. At the midpoints of the seven houses mentioned we find Mars in the twelfth house, natal Uranus in the eleventh and the natal Ascendant in the first house. Just before or just after - not more than a day's difference - such positions, in addition to Pluto, are occupied by natal Pluto (and the natal MC next to it) in the tenth, Mercury in the seventh, and natal Saturn in the eighth.

Even if we recalculate the degrees of the ranges of these houses, a strong correlation in time is found, and the difference of a few degrees in the positions of the planets from the centre of the house only speaks to the fact that not all the problems happened on exactly the same day. After all, in a lunar return, each degree counts for about one day of that month. With so many problems piling up, Scorpio's resilient nature could no longer cope with the burden.

PLANETARY RETURNS

As mentioned earlier, similar to the solar and lunar returns, for the purpose of prediction and better insight we can also calculate the returns of other bodies, the planets, returning to their birth positions with their own frequencies or orbital periods. Mercury and Venus do this about once a year, Mars every two years, Jupiter every twelve years and so on. An astrological chart calculated for the exact moment of a planet's return to its exact natal position, like any other astrological chart, shows the picture and quality of the moment, and thus the potential for the period, covered by the return chart. The chart of Mercury's return will therefore show the potential, lasting until Mercury's next return to the same position, the starting point, after a year or so, as will the chart of Venus' return and all the other returns in the series up to Pluto.

If a planet happens to be placed in the same astrological house so in the return as in the natal chart, the respective planetary return is of great significance and should be paid special attention. It goes without saying that this is more likely to be a Mercury or a Venus return, since the number of these and solar returns in our lifetime is approximately equal; with other planets, however, - in accordance with their distance from the Sun or their place in the planetary line – this chance is getting less and less frequent.

Returns of personal planets

Planetary returns are particularly useful, when someone is interested in the dynamics of the planets or themes in life, represented ("ruled") by the respective planets.

Mercury's return will therefore be helpful with topics 'covered' by this planet, like relocating, travelling, signing a contract, making an important decision or buying a car. It may also be useful for anyone in a Mercury related profession or mission. Its association with mental processes offers a useful guidance within the year, represented by Mercury return chart. As with the Solar Return, the positions of the Sun and Moon are important, however, Mercury is the planet to pay most attention to, both in terms of its house position in the return horoscope, as well as its aspects with the other elements of the horoscope.

Venus' return reflects our contacts and relationships with others, harmony, love, marriage, as well as money and value. The house Venus is placed in indicates the general nature of the respective activities over the time, represented by such a horoscope, i.e. about a year. In a Venus return horoscope of a male person, Venus also indicates the quality of relationships with women of importance at the time, its position showing a favourable or a negative development in the respective sphere.

Mars' return refers to about a two-year period of our lives, and can be regarded as a kind of beginning of a new cycle of personal initiative and enterprise. In such a horoscope it is possible to notice and pursue new goals and achievements in the respective period, as well as to discover a favourable period for activities, requiring great effort, energy and breakthrough, and leading to success. A Mars return may also help us avoid dates of potential pitfalls or defeats. In a woman's horoscope, the position of Mars shows the quality and possible development of relationships with the opposite sex during the time, represented by the chart. Understandably, the Mars return chart is also most important for people with strong and significant Mars in their natal horoscope.

Slow planets returns

The return of **Jupiter** is an astrological chart, representing a period of twelve full years. Since this planet, as the 'Great Benefic', denotes growth, good fortune, protection, along with all things auspicious and pleasant, this is what we search for in its return chart, especially through transits and progressions. However, the question arises as to whether we are (can be) even aware of such a long cycle in our normal life – as compared to the past ones – to notice the differences. We experience and encounter many different variables over such a long period of time, including ageing, which cannot be overlooked. On the other hand, it is also true that, due to the time difference between Jupiter returns, it is possible to observe differences between the charts which, also because of our age, are perceived quite differently from the previous ones, even if they (might) be similar to some extent.

The principle is similar to the "ingress" charts of Jupiter and other slow planets into the signs - even if it is the same position, e.g. Jupiter entering Aries, each chart is different from the previous one, meaning that each successive entry brings a significantly different Jupiter energy.

These dilemmas are even more perceptible in the next return, that of **Saturn**. A Saturn return represents a period of 29 and a half years, which means we can experience a maximum of three in a lifetime, taking life expectancy into account, and the specificities of the first one versus the second one, of the second one versus the first and the third one, and of the third one in relation to the first two, are so different that comparison is almost out of question. The **Uranus**, **Neptune** and **Pluto** returns already illustrate periods beyond our lifetime.

There are, however, some astrologers, who also use these returns. Namely, each planetary return can be divided into 12 parts, representing the time that planet takes to travel through exactly one sign. If each planetary return represents 'the return of a planet to the exact position it occupies in the natal horoscope', i.e. a 360-degree journey, this makes twelve segments of 30 degrees. For example: Jupiter, placed in the natal horoscope at, say, 5°15 Taurus, will reach 5°15 Gemini within a year or so, then 5°15 Cancer, and so on. Saturn will take about two and a half years to make the same journey, and Mercury an average of one month, not taking into account the specificity of its motion, i.e. the great differences in the speed of its direct and the three retrograde motions within one year.

Triple returns

There is another peculiar feature about planetary return horoscopes. The Sun and Moon, which are in fact luminaries and not planets, are the only two bodies in the horoscope that are never retrograde, i.e., seemingly going backwards. With all the other planets there are retrograde phases and periods, which are quite different from each other for the personal planets, although for the more distant planets - from Jupiter onwards - the retrograde periods are becoming more and more predictable. A planet turns retrograde roughly somewhere in a separating trine with the Sun, ending the phase at the approximate formation of the next, this time applying trine with the Sun, meaning in simplified terms that slow planets are retrograde when on the other side of the sky from the Sun, i.e. in opposition to it with respect to the Earth's position. This is true for planets from Mars onwards, while Mercury and Venus are closer to the Sun than the Earth, and their retrograde positions can therefore not be determined in this way.

A Mercury's retrograde path may start close to its natal position, and after a few days past that position, it may return back along the ecliptic, and then later pass the same point again; each time we make a return chart from the retrograde passage, which means that we can have as many as three Mercury retrogrades in a given year. Similar rule applies to other planetary retrogrades.

With Mercury, the horoscope of such an "intermediate" planetary return may only last a few weeks or even just a few days. Then the next one comes along, and yet all these charts have the validity of all horoscopes, representing a moment in time and the potentials of the following period. No published studies of such cases are known, however, it would be interesting to examine the applicability of such charts in real life, as well as in astrological practice.

VIII
ABOUT OUR PERCEPTION

If in the previous chapters we have managed to master the division of time within a year, it is necessary at this point to clarify a crucially important principle, namely the perception of events.

WHEN DID FATHER DIE?

One of the best possible examples is the case of a lady from a small Slovenian town, with whom we were struggling to regain her inner balance after a very difficult and turbulent year. In reviewing the events of the past year, it was clear that one of the main shocks was the death of her father. As usually, we must first examine the natal horoscope, where relationship with the father is shown as strong, solid and positive. According to the lady's own statement - confirmation - it was a relationship that was more than just a father-daughter relationship. She said that for her (especially after her mother's death a few years ago - "it didn't affect me too much...") he was almost a god, and after her divorce the father, the client and her daughter lived together very harmoniously.

Then her father passed away suddenly from a heart attack, in his sleep. If Pluto is found in the fourth house in the yearly horoscope, squaring Jupiter, the co-ruler of that same fourth house, "destruction at home, destruction of the father" can be expected. In addition, Uranus in the sixth house and Mars in the eighth house are also active at the same time of the year, the simultaneity of several elements indicating the importance of that particular part of the year!

So far so good, except for the timing. Given the position of this planet, it is clear without further calculation that the father died sometime in July, but much to my surprise she stated that he died "exactly one year ago", i.e. 26 April 2004!!! (The consultation took place in April 2005.) That points to an incredible three-month lag, meaning, according to the principle of proportionality, an inaccurate birth time, about one-quarter of the house making a difference of about half an hour in the time of birth. But after a few more questions to check the time of birth, which up to that point in the conversation had not been disputed, it turned out that the time of birth was fine. So what to say now?

The astrologer was left with only one possible explanation of the discrepancy: *"Is it possible that you only **grasped the fact** of your father's death three months later?!"* A completely nonsensical and bizarre question, followed by astonishment, brief silence and confirmation. And by my astonishment too. Because had they not lived together, it might still have been plausible somehow, because even the death of the closest person, in certain circumstances, may not be related immediately, but sometime later; however, not grasping the death of someone very close, who, until recently, had been part of your everyday life, is truly unusual.

SOLAR RETURNS

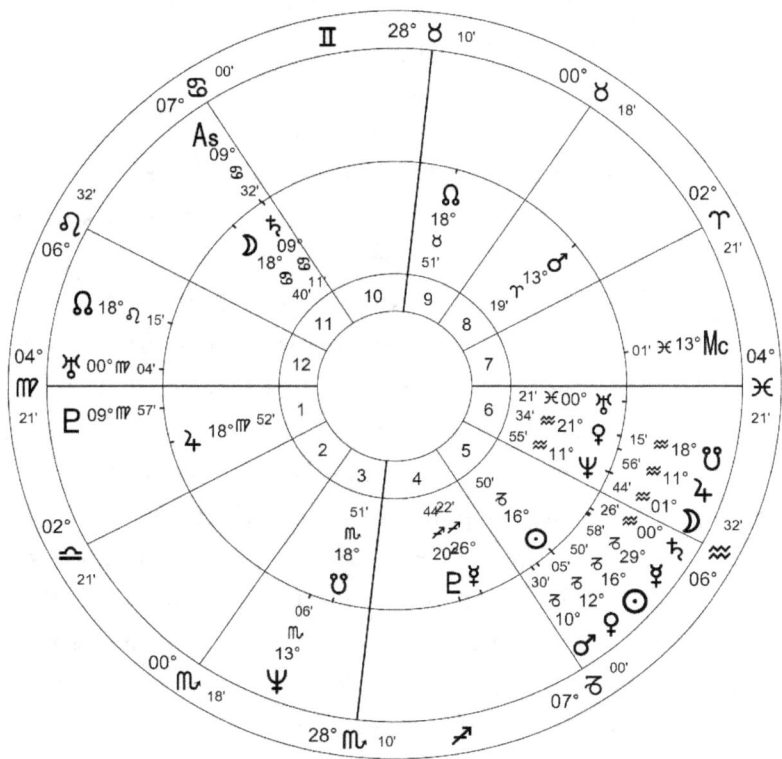

But that is exactly what happened: *"For three months I had been hoping he would turn up eventually and we would be together again."* Three months later, the bitter realization finally came through for my client, and her neighbours, also acquaintances of mine, later confirmed that she "had been walking around like a zombie for at least another month or two", and that she had to be literally reminded to wake up at cash register in the shop if she was not to make a big queue.

Pluto in the fourth house does indeed indicate "the destruction of the home" or - according to the Croatian astrologer Mile Dupor

– "the demolition of the hearth", but in the horoscope only the **perception** of this demolition, not the **demolition itself is visible!** The two do not have to coincide in terms of time. The long time lag in the perception of an event is easy to explain by examining the cases, in which a person was not present at the event itself, and the information about it only came later. This also explains the "classic" response of the client, when trying to identify an event or a situation, namely *"Yes, it may well have happened, but I just don't know"*. Such an evasion is simply invalid - for if something is written in the horoscope, the respective person has to be aware of it, one way or another! If not, it cannot be indicated in the horoscope. **A horoscope therefore is not a picture of an activity, but rather a picture of our perception of what is happening, there being a big and essential difference between the two.**

However, the 'I don't know' cannot be dealt with in such a short and simple way.

WHAT IS INDICATED

Verification is also an important part of astrological consultation work. We ask the client about some key recent events and try to identify them in terms of time, on the one hand to make sure the horoscope and the birth time can be relied upon, and on the other hand to understand the client and his/her way of reasoning and perception to the extent, that the astrological symbolism can be presented to him/her in an accessible and understandable manner. For this reason alone, when preparing the calculations for an astrological consultation, I usually also make at least one solar return horoscope for the year before the date of the consultation. As already mentioned, it is possible to figure out feelings from a solar return horoscope, while a trend or development of activity is visible in a series of solar returns. Indications in a horoscope show individual and related significant events or feelings, helping us to construct a story on the one hand, as well as a possible further development on the other.

And what is 'important'? Understandably, this depends mostly on the individual person and his/her character. Yet let us try to imagine this dilemma simplistically: an event of boilt over milk on 15 March last year is unlikely to be remembered for more than a day or two, or not even that, being of no importance. But if that boiling milk was accidentally spilled on our child and we ended up in the emergency unit, we are unlikely to ever forget it, or at least not for a very long time. And yet, in both cases, technically speaking, it is just - 'milk that boilt over'.

Notes in a horoscope, in this case in a solar return chart, bear a certain meaning, relating to something that was important to us at the time, defined by the respective position. Important **at the time**, however, but not necessarily important later or at the time of astrological consultation. So it is possible to come across a note, and the client will not be able to remember the event associated with it after a time, however hard he tries.

In such a case, we simply move on to other points in the horoscope, and experience shows the meaning of the note to usually be clarified in the course of the consultation. Moreover, if in a particular horoscope several planets are located in the same time period, even if **not mutually aspected**, this is certainly the most active and important period in the respective year.

Every point, element, planet, aspect, etc. in the horoscope can be interpreted in several ways, and this is evident in every astrology textbook. The ninth house, for example, signifies the mind, knowledge, study, foreign affairs, tourism, religion, languages, documents, legal and administrative procedures - but which of these fields is most important or useful for the client in a given situation? In an astrological consultation, we need to know the client well enough to foresee his **understanding of our interpretation.**

DISINTEGRATION OF A SYSTEM (SYSTEM FALLING APART)

Let us look at another example of a specific position in yearly horoscope that is strongly linked to the dilemma of perception, and has been demonstrated and verified so many times in practice that it can be regarded as a 'definite indicator', adding to consultation a kind of 'package' of information of great utility. It is Pluto's position in the twelfth house.

One of Pluto's roles in horoscope is that of "destroyer", and the 12th house signifies something hidden and unpleasant, as well as helplessness, depression or nervousness, all very unpleasant states in general. Neptune's influence will add to this ambiguity, being associated with dreams, the subconscious, the fluid that fills our brain, and consequently with the possibility of influencing our thoughts and mental state. Pluto's position in the twelfth house could thus be called, in a nutshell, "the breakdown of the emotional system". Often it is associated with a feeling of "the world having been turned upside down" or "it's over now, I can't take it anymore!", and not infrequently a person with such a position will suffer from a collapse or nervous breakdown. Helplessness grows to such an extent that one can no longer control situation, consequently "breaking down".

The situation is most like a glass overflowing, waiting for just one more drop to spill over. But with the glass being full and overflowing, what is its content? Some problem that cannot be handled, being too big, too heavy for us? Usually not, except perhaps in case of some phenomenon or event we have been fearing for a long time (smouldering in the background, associated with fear or helplessness, building-up of tension = Pluto), but when it eventually does happen, we are crashed. As a rule, such a collapse is not due to a single cause or reason, but rather to an accumulation of various, more or less unrelated components, each contributing to the severity of the period, and once too many have accumulated, one has had enough and cannot take it anymore.

As an illustration, here is an example from my own long-standing practice, which made me become aware for the first time of the seriousness of the situation. It was related to me by a client, a primary school secretary. Her 'glass' contained a lot of activity and tension after a recent divorce (with adjusting to new conditions, paperwork and discomforting contacts with hitherto mutual friends), with a sick daughter, who had to be taken care of in the morning before going to work, a broken-down car, unexpectedly held up at the garage the previous day, as well as with the associated anxiety about the likely higher-than-expected cost, the need to get up earlier to walk to work, and, last but not least, an unpleasant fall on a frozen pavement just after leaving the house. So she arrived to work with a severe pain in her lower back, which would have kept many people at home, and which made it very difficult for her to work, mostly in a sitting position. When one of the senior pupils came in during the main break, reproaching her with a pubescent insensitivity of not having been capable of handling her work as she "could not even manage to put paper into the photocopier", my client got up and walked over to the paper cupboard, intending to hit the girl's head with a pile of paper, but it never happened - she fainted and passed out, hit her head on the parquet floor and woke up in hospital, where she was kept under observation for a few weeks on suspicion of a fractured skull.

Was that pile of paper really so important? Not at all, and neither was the reproach of a teenage girl, which she was used to, having been part of her job, besides she was no beginner. It was just the 'last drop in the jar'! The fall on the tailbone and the pain, the car, the expenses, the lack of sleep, the daughter's illness, etc., had been in the jar before, having been piling up for a long time, with the tension mounting. In fact, it was only a matter of time for her to break down, with practically anything as a trigger, perhaps a remark or criticism from a superior, a headmistress, a phone falling on the floor, or some trivial argument with a colleague… It could have been just the realization of having forgotten something crucially important at home, or that she now had to walk the same route twice, ask someone for a car, or for a lift home and back, or maybe a short heartache… In another case from my archives, a similar collapse was triggered by the realization that, at the very moment when a phone call would have saved her from a very distressing situation, my client's phone battery was empty…

Another client described the situation as follows: *"It was in the early 1950s, in the time of the 'Trieste crisis'. We were all afraid of a new war, having had no phone, TV or internet to find out more, information was scarce. I was in the ninth month of a very difficult pregnancy, having been grateful to God for my husband, who did everything for me, including the house and barn work, all the things I had been doing for all those years. I could not even bend down and touch the floor. Then there was a knock on the window, it was half past two in the morning, and they took him away almost without a word. To the front. He only had time for one hug and the promise 'to be back soon'. And me in bed, with two small children and three heads of cattle, which required a lot of effort! I thought I would go mad. No information, no call, no letter, all through the time of crisis... If I hadn't been so highly pregnant, I would have killed myself, I would have done it out of sheer helplessness, so bad it was..."*

The most unpleasant but also the most important thing for understanding this placement, is the definite inner feeling about many things in the accumulated pile that could have been dealt with before, perhaps days or years ago; we did not do it, either because we could not, did not dare to, did not feel like it, forgot, or whatever – Pluto's 'invoice' is always very unpleasant, and the recognition comes in a flash of a single moment!

Not every Pluto is Pluto

Is every placement of Pluto in the twelfth house of the solar return horoscope so hard and unbearable, leading to a 'breakdown of the system'? Of course not, but at this point some differences have to be defined first. The first thing to note is, whether it is a solar Pluto (inner circle) or a natal Pluto (outer circle), and the second is the specific nature of Pluto's placement in the solar twelfth house, as well as it's natal position.

The first rule of prediction is that a planet favourably placed in a natal chart cannot be very unfavourably placed in another chart, and vice versa. This means that a Pluto – from which we cannot as a rule expect anything really positive – having "only good aspects" in the natal horoscope, will not bring an extremely unpleasant situation in the yearly horoscope, even if negatively aspected with the solar planets. And vice versa, of course.

Let us have a look at an option, where unfavourably placed Pluto, is not very "badly" aspected to the rest of the horoscope. Pluto in the natal twelfth house can also signify a person, who works from behind the scenes, has hidden agendas, lobbies, spies, pressures and terrorises, is very powerful 'underneath', capable of achieving things that no one else, including himself, could achieve otherwise. In the yearly horoscope things are no different. Here again the pressure can be enormous, but a person, who finds himself under pressure or decides to 'do something', does not act in the usual or direct way. In such a case, the popular phrase "beggars cannot be choosers" comes into play – a person with such a horoscope probably doesn't like acting in the manner of a terrorist or blackmailer, however, if there's no other way... Pluto shows his extreme nature here too - the 'all or nothing' or 'it's either you or me' principle doesn't allow for much choice. The result is, on the one hand, the feeling that we had had no choice, followed by the realization that we escaped the worst predicament, and lastly a very bitter taste in the mouth, caused by the moves or procedures we made (had to make!), if we were to get out of trouble; now it is up to those around us to resent it or not... Pluto is certainly not some kind person one would invite for a coffee.

A different situation arises with **natal** Pluto in the 12th house of the yearly horoscope, i.e. the one that will be marked on the outer circle of the horoscope. Pluto is of course still considered unpleasant, so we can anticipate strong pressures and sleepless nights, however, the difference is that these are 'outside' pressures (not necessarily any less), meaning that the respective person will not feel that she/he deserves these problems for having allowed too many things to pile up. On the contrary, there may be pressure from one direction or from one person or thing, but this being Pluto, the pressure makes it impossible to breathe, making the person feel trapped and helpless.

Again, it depends on Pluto's aspects to the solar planets, as well as on its placement and aspects in the natal chart, whether or how one will get out of trouble. What is more, Pluto will almost certainly let one know at some point that he/she had a "once before" opportunity to resolve the present tension, but failed to do it...

And what can be done?

What are we to do with such a position in a solar return horoscope? Is it a signal for panic, a prediction of helplessness and 'system collapse', or do we have other, milder options for resolution? How do we deal with this Pluto?

In short: make sure the glass is not overfilled. Just this. Very simply put, and that is enough. It does not, however, mean it is easy to achieve. To do that, let us try to solve at least some of the accumulated problems, some of those little annoying ones that can be solved easily, without too much trouble, except that we need some time (our usual excuse, which Pluto doesn't understand...). The fewer drops (!) in a glass, the less likely a collapse due to some photocopy paper or a dead mobile phone battery. If we have the time to prepare in this way, that is, if we are aware of the culmination of such a feeling months in advance, we have the opportunity to do more, perhaps solve one of the major problems in the pile, perhaps even one that has not been dealt with for a decade...

The same advice can be given to a person with such a horoscope, yet we must not be too merciful, because Pluto never is. A little tact and an ethical approach will be enough, and when "that very" time comes, the client will be most grateful. As a practising astrologer, I have at least a dozen written heartfelt thank-you notes in my inbox for timely information about such a Pluto position, along with descriptions of examples of what had accumulated and how it had unravelled. The astrologer's information did not solve the problem itself, however, it did bring the opportunity to 'do something', and what is better for the (alerted) person than the feeling of having saved herself out of trouble on her own?

Now that we know how to determine the time frame in the solar horoscope when a potential will be most pronounced, we can also do this with Pluto in the 12th house. This way, our clients can get information in time, allowing for a wider scope of action. But to give an example, a woman came for a consultation only three weeks (!) before a calculated and predicted collapse. After a carefully devised 'tactic' she left, and, a good month later, reported on how things had gone, and how she had successfully extricated herself from a situation that could have had much worse consequences for her, as she put it... It is true, however, that in the meantime she had missed out on minor tasks and even absented herself from work in order to be better prepared.

Finally, Pluto in the 12th house is by no means the only indicator of a collapse or nervous breakdown. Depending on our perceptions and priorities, there may be other positions, or a confluence of influences from different planets, resulting in similar outcomes. However, in my experience over the years, it is Pluto's position in the twelfth house of the solar return that requires special attention and vigilance, both in preparation for the time of its greatest impact, as well as in the consultation itself.

IX

SOLAR RETURNS IN PRACTICE: MARGARET THATCHER'S BATTLES AND VICTORIES

So far we have had a good look at the solar return technique explanation and interpretation, its advantages (with the help of examples) and especially the ease with which a solar chart can be read and interpreted. However, a better overview, based on a single example, can be gained from the case of a person, sufficiently well known and at the same time sufficiently interesting to derive a number of indicators from her solar charts survey. There are several reasons for the choice of the former British Prime Minister Margaret Thatcher for this attempt at astrological life-tracking. First of all, being a sufficiently well-known figure, she needs no special introduction, her life having been known to a wide range of people at the time of her political activity.

Horoscope of a great career

As always, her birth horoscope will serve as a starting point for further work with the solar charts. Margaret Thatcher was born on 13 October 1925 in Grantham, near London, UK. The birth time, recorded as '9.00', immediately raises doubts as to its accuracy, since it is quite possible that it was recorded more or less approximately, as was the habit in those times. But this is not the case - working with her horoscope, and in particular using various astrological techniques, quickly proves her birth time accurate enough to be considered reliable. If that is the case, it can be trusted that some of the notable events from her CV will also show up in her solar charts as accurately as possible, at the very points where we would expect them to be, once we have mastered the interpretation of solar charts.

A Libra Sun is expected to be diplomatic, hesitant and weighing long before making decisions, to be reluctant to fight and conciliatory, but history proved this not to have been the case. Prime Minister Thatcher showed more firmness than many a statesman during her reign, which can first be confirmed by the Scorpio Ascendant , then by the position of Pluto, ruler of the Ascendant, in the ninth house (this planet not allowing for others to assert their opinions in her presence), and especially in the stellium of Mars, Sun and Mercury at the end of the eleventh and beginning of the twelfth house, which together, in addition to everything else, form a T-square with Pluto, ruler of the Ascendant, on the one hand, and Jupiter on the other. Mars, as the co-ruler of the Ascendant, gives this configuration a special character, being extremely energetic, combative and relentless in this case.

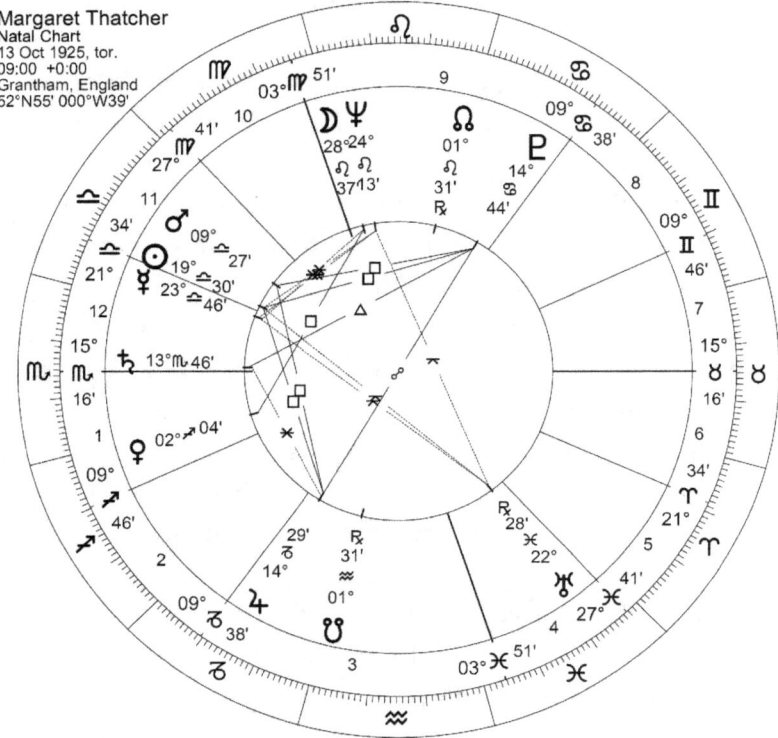

The Moon is placed at the very end of the ninth house in her birth chart; as the highest, culminating planet it further highlights the ambitious character of Mrs Thatcher's. "She was somehow predestined for a great career", and "it felt good to be successful", one might say.

Mars in Libra does speak of a strong tendency to compromise, but its aspects, including its conjunction with the Sun, make it clear that there were mainly "compromises by force" in her career, Mrs Thatcher having been rather adept at imposing her will on her surroundings. Her inner strength is also well attested by the powerful Pluto-Saturn trine, making her capable of long, persistent and relentless work until the task had finally been accomplished.

Saturn is also conjunct the Ascendant, increasing her ambition, and contributing to her cold temper and sharp tongue (again a Pluto connection).

At a first glance, a greater emphasis on the left, i.e. eastern, side of her horoscope can be noticed, making her "Me" pole much stronger than her "others" pole; besides, most of the planets are placed in the upper half of her natal chart, which - in addition to some of the indicators already mentioned - suggests high ambition. As a footnote, this is the horoscope of a person, who is certainly not bound to live an ordinary life, filled with household duties and trivial socializing. Not Margaret Thatcher, who had a more important, "higher" role in life.

Margaret Roberts and Margaret Thatcher

But that is not how things seemed at the beginning. She was born Margaret Hilda Roberts into a family of a barber and a seamstress, both of whom were entrepreneurial enough to provide prosperity for the family. Her father was a local politician, who was encouraging Margaret, the younger of the daughters, to read and keep up to date with the world affairs, to be a leader, not some subservient person in service of others. If father is indicated in a birth chart by the Sun on the one side, and the fourth house on the other, Margaret obviously greatly admired him, while he was also very important for her future (the Sun is in the eleventh house).

Her interest in politics started early, having become President of the Oxford University Conservative Association at the age of 21. It is less known, however, that her primary education was chemistry. Early on she seemed to have been inclined to research, a profession not associated with a woman, living in the post-war England. Astrologically, such a possibility is suggested by the emphasis on the twelfth house, the house of research: the Mars-empowered Sun is on the very cusp of this house, along with Mercury, and Venus; the latter, ruling the twelfth house, is in the first house (research - my identification). But things have turned out differently. For the potentials of this horoscope, the fame of the explorer would obviously be insufficient.

Soon after her recruitment, she was promoted to assistant manager of the company, and in 1949 ran for the first time for a seat in Parliament. She was unsuccessful, but still received 36% of the total vote.

Her first election was not only important as her first political test, but also as a major turning point in her life; she met Denis Thatcher, a businessman and millionaire, whom she soon began dating and married in 1951.

Wedding

Margaret Thatcher met her husband-to-be at an event in February 1949, while he was still in the midst of his first, unsuccessful marriage of only a few years, in which he had no children. They married on 13 December 1951. Let us therefore have a look at the solar horoscope of the wedding year, i.e. the horoscope for her birthday in 1951, also covering December 1951, a time about two months after the birthday. (For convenience, we will also refer to this solar horoscope as the "1951/52 solar".)

SOLAR RETURNS IN PRACTICE: MARGARET THATCHER'S BATTLES AND VICTORIES

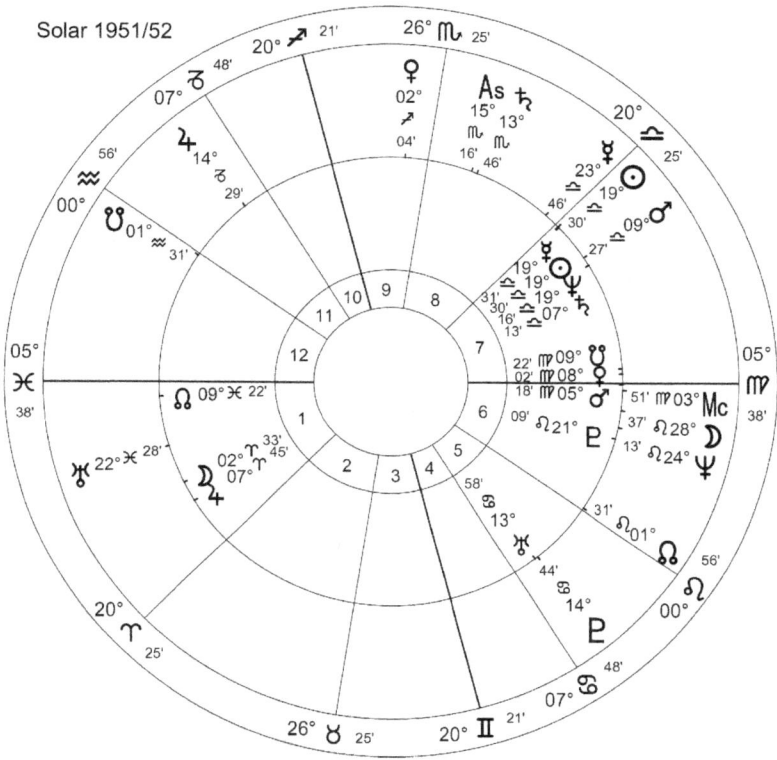

Solar 1951/52

The Ascendant of this solar return is in Pisces, and the vast majority of planets are placed on the right side of the chart, in the sphere, where others prevail - at least for this very year. There are two Ascendant rulers; Jupiter in the first house close to the Moon, and Neptune in a much more illustrative position - in the seventh house (marriage) and also in a close conjunction with the Sun (self) and Mercury (change); the latter also rules the 4th (home) and the 7th (marriage) houses in this chart.

As for the time of wedding in this chart, the stellium points to a time just – perhaps a few days – after birthday, but a closer look shows a different time, about a month or two after birthday; the Moon's node, namely, is in the first house, natal Venus, ruler of the

Sun and also of the seventh house in the birth horoscope, is in the ninth house (legal matters), and last but not least natal Mercury, ruler of the tenth house (status) and the eleventh house (the eleventh house of the birth horoscope), is also important. The positions correspond perfectly to 13 December.

Two such different indicators can only be confusing if we are unable to understand marriage in its full complexity, taking into account the circumstances and, above all, the times, quite different from the present day transitory values. Preparations, decisions, a society full of protocols and complicated relationships, but also doubts (the Ascendant in Pisces can refer either to idealization or to doubt, ambiguity). Had we known what was going on in those months on the margins of the main story, the various indicators would suddenly become meaningful.

It is also well worth checking the previous solar return, a time when decisions and preparations were taking place. This horoscope does not exactly show a wedding, however, it is interesting from another point of view. Venus as the ruler of the seventh house is placed in the tenth house in that horoscope - "marriage as a matter of status"? There could be a small disappointment here. For all the indications Margaret Roberts was more concerned with the status, gained through marriage, than with marital bliss, happiness, children and the like. But let us not be distracted by the well known later image of Denis Thatcher as the "silent husband in the background". **At the time of the wedding**, and this is important at the moment, Denis Thatcher was a millionaire, while Margaret was "only" a very ambitious steel company employee and unsuccessful parliamentary candidate, from a family that did not even come close to Thatcher's knees! It is only natural for such a marriage to be reflected in the tenth house and not just in the seventh... Her thinking at the time was probably along the lines of "I'll do better next time with this marriage!"

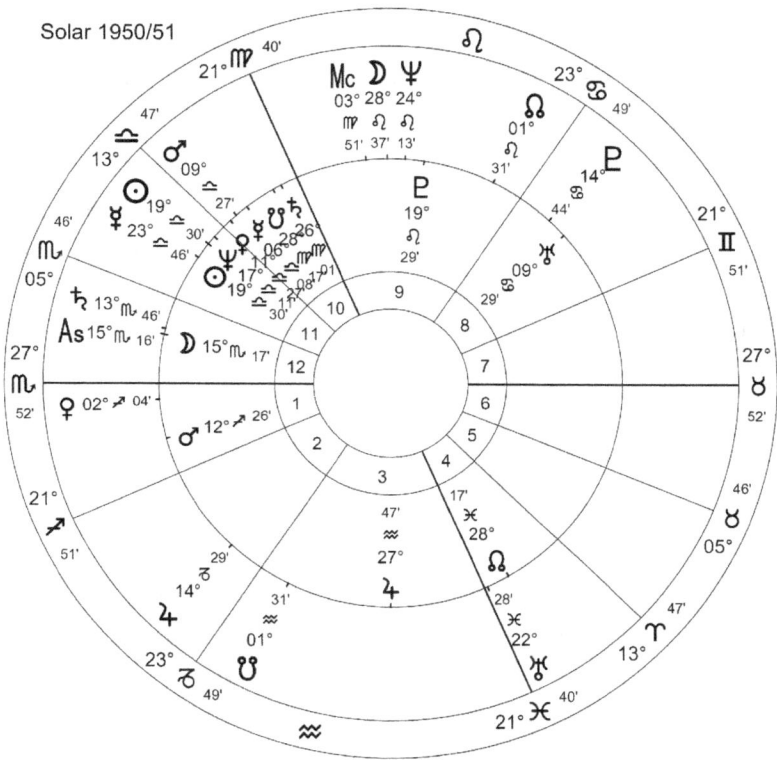

There are two other important points of the 1950/51 solar return. Saturn at the cusp of the tenth house points to a sense of failure in achieving ambitions, and the South Node immediately after to a feeling that "perhaps it was better this way". Much more interesting, however, is the solar return twelfth house this time. Already very pronounced in her natal horoscope, it is even stronger in this solar return: the Moon is practically on natal Saturn (which is also in the 12th house, so this is a repeated position!), also squaring Pluto. Could Margaret have been pervaded by very strong and mixed feelings simultaneously? For example, that this marriage could bring her success in life, along with a prison? Could it have been her mother pressurizing her, having sensed 'a perfect opportunity for her daughter'? Pluto pressing down on this poor Moon (Margaret suffered from the worst psychic pressure sometime in March-April 1951) is after all in the 9th house; - 9th house = expansion, Pluto = destruction → "the end of freedom for me"?! Could that have been the case? Possibly, but fortunately the Moon has a "redemptive" sextile to natal Jupiter, this time placed in her 2nd house - money and a sense of self-worth eventually prevailed.

This suddenly shows Margaret Thatcher in a very different light, however, it was not yet known at the time that in future she would be known as the "Iron Lady", determined to go on at any cost...

Margaret as a mother

The next important event was the birth of her children. The first information available was sparse: "In 1953, the Thatchers had twins, named Mark and Carol". This could be referred to by two solar returns, 1952/53 and 1953/54. The former contains quite a few indications of childbirth or at least pregnancy (or rather first motherhood, which for most women means not "just" having a baby, but also a change of status): the Ascendant is placed in Cancer, the fourth house is quite wide as well as highly pronounced, with vast majority of planets placed in the lower half - this was undoubtedly not a year of progress and career, but rather a year of stagnation; even the Moon, representing motherhood, the leading planet of this annual chart, is placed in the second house, the house of value.

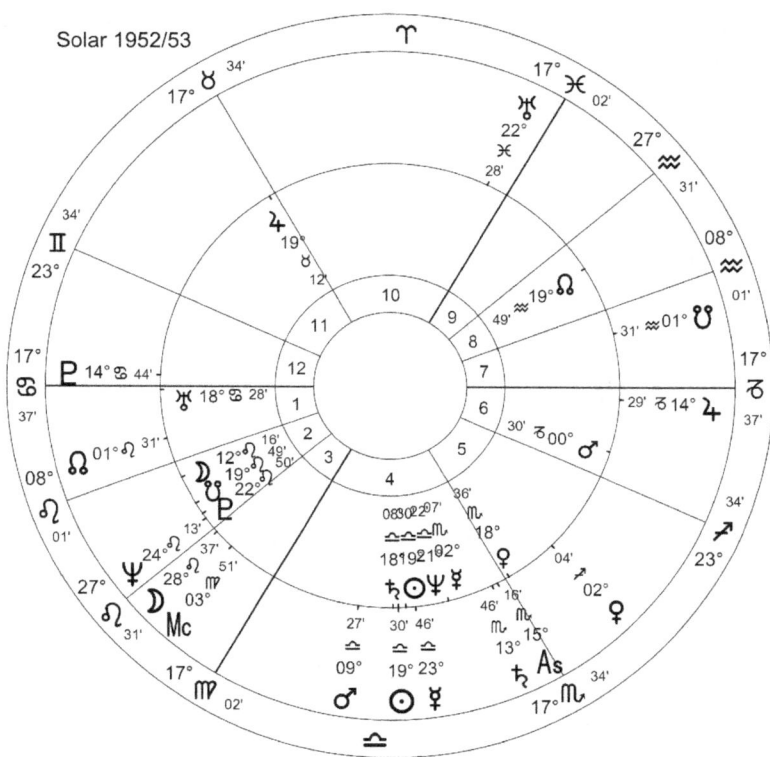

The next chart, Solar return 1953/54, brings less indications of pregnancy and childbirth, although not lacking them either. The Ascendant of this chart is in Virgo as one of the Ascendant signs usually related to motherhood [10]. The ruler of the chart Mercury is at the very end of the 2nd house, along with natal Saturn, Mars as ruler of the 3rd and 8th houses is conjunct Ascendant. The main indicator of motherhood, the Moon, is here in Sagittarius and in the

---

10 As already noted, the year of childbirth or change in the number of children is often marked by Ascendant in Cancer, Gemini or Virgo, and in the case of a family expansion (i.e. not the first child), Sagittarius can also be found there. In the two years observed on this occasion, two of these signs follow each other, first Cancer and then Virgo in 1953/54.

4th house (home), but only as the ruler of the 11th house, which has nothing to do with children. This is not exactly a chart of motherhood, meaning there were probably other important matters taking place in that very year. The children were born on 15 August 1953.

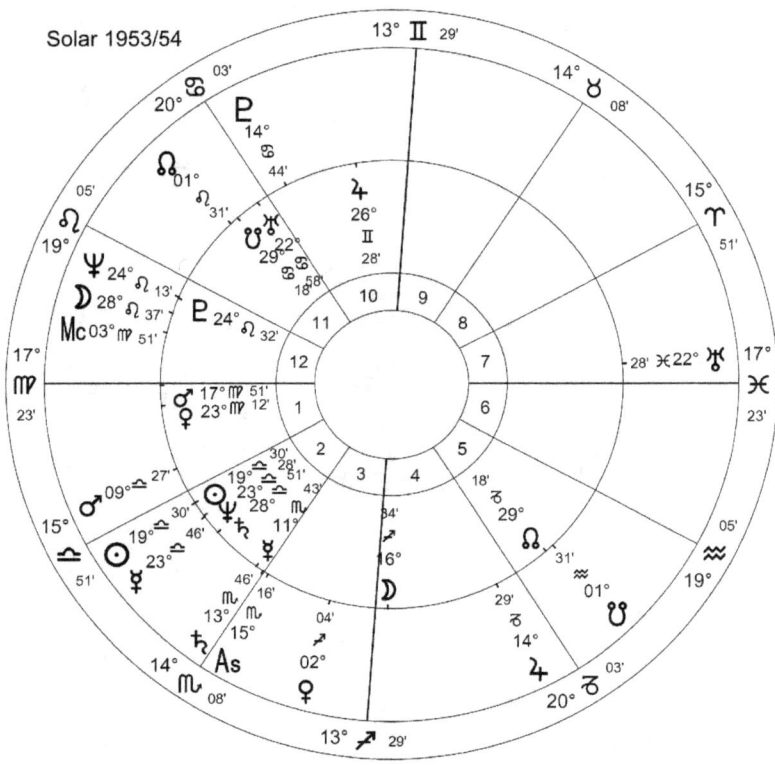

Solar 1953/54

Another interesting fact from this time is that "four months after the birth of her children, Margaret passed the bar exam". Shocking – with such 'an important exam not being easy to pass, she must have studied very hard - during high pregnancy?! What is more, it is also known that she sent the application form straight from the maternity hospital after the birth of the twins...

For Margaret, the bar exam was a springboard to career, so this kind of success should also be reflected in her horoscope. Jupiter in the tenth house (career, success) or in the ninth house (studies, law) would be most appropriate for this. And lo and behold, we find Jupiter right there in the tenth house in 1953/54, but at a time, which cannot be defined as "4 months after the birth of the children", but rather later, sometime in late February or early March 1954.

However, at the time corresponding to December 1953, Venus is found in the first house (the self) as the ruler of the ninth house (the studies), along with Sun and Neptune in the second house, pointing to the value that the young mother had achieved in this way. After all, this also gave Denis an heir, an additional, perhaps silent recognition for Margaret. Jupiter in a "delayed" position, instead of being a direct indication of having passed the exam, could thus also be interpreted as the "feeling of having achieved something great".

However, there is interesting activity in the solar return shortly after the birth. We quoted the information that Margaret passed her bar exam "four months after giving birth". Pluto in the 12th house is indicative of extreme stress, of a glass full to the top, when water starts spilling over, simply because too many things have accumulated for the tension to be contained. Pluto is placed two and a half months after the birthday, sometime around New Year. Pluto here rules the third house, the house of learning and also of relationships. But just slightly preceding Pluto there is natal Neptune, ruler of the fourth and fifth houses. Neptune in that place may indicate "something being wrong" with children and home, for example, that these two concepts may have been idealistic, illusive, vague and undefined. There is a self-evident possibility that Margaret probably neglected her motherhood, and that it was getting on her nerves, having had higher goals that were not compliant with her parental role.

Uranus in the 11th house, placed earlier, speaks of her decision to seize the "opportunity for the future". The aspects of Uranus are particularly interesting: it is unfavourably aspected to Neptune in the second house, which conjuncts natal Mercury, ruler of the 8th, 10th and 11th natal houses. On the other hand, Uranus has two good aspects, a sextile with the aforementioned Venus in the first house and an exact trine with natal Uranus, placed in the seventh house.

This Uranus is also adequately placed in the seventh house regarding time. Since as a natal planet it is placed in the outer circle, indicating "other people, external events and circumstances", it can be concluded that Denis did not agree with his wife's plans, which, in turn, generated a marital crisis. Denis, however, did not have much chance in his efforts to bring the family into a "normal", i.e. expected and traditional state, as shown by Margaret's Mars, which in this year's horoscope is placed on the Ascendant itself, indicating a year of struggle, hard work, tension and battle. Margaret simply "had" to be a winner this year!

Election to Parliament

The next event to examine with the help of solar returns is Margaret's election to Parliament, which took place on 8 October 1959. Here again two successive solar returns have to be considered, the event falling right on the time boundary between the two, and it can be said that Margaret's election to Parliament in her second attempt was almost literally "her birthday present".

The first of the two returns, the one for 1958/59, at the very end of which Margaret was elected, and which is supposed to show the feelings during the campaign, is very interesting. As in the natal horoscope, Ascendant is again in Scorpio, indicating "a year of crucial significance for the later life". Pluto, ruler of the Ascendant and horoscope, is placed high in the tenth house as the culminating planet, meaning that - "no one can throw me out of the saddle now".

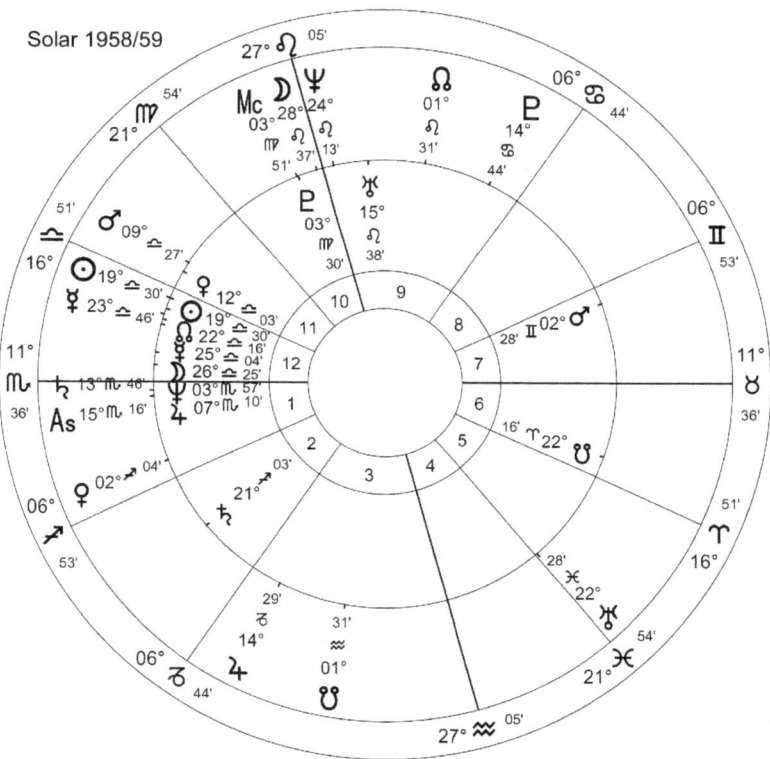

The activities are evidenced by the strongly highlighted first house on the one side and Mars in the seventh house on the other. It is a well known fact that every political campaign is difficult and dirty, so Mars (known opponents) is not surprising. But the full 12th house is much more interesting. Had Margaret Thatcher not had the same house highlighted in her natal horoscope, this year could be described as *"difficult, unhappy and full of confusion"*, but in this case these are the fields innately *"familiar and favourable"*. Mercury plays a major role here, being almost in the same place as in the natal horoscope, quite in accordance with the established rule of a repeated position in prediction. Mercury here speaks of the thorough consideration and time Margaret took for her further moves. Neptune, placed in the same house and at a slightly later time, shows that at a certain time (sometime in early July) she was overwhelmed by a feeling of uncertainty, ambiguity, illusion, perhaps even indecision, and certainly not quite sure of success. Jupiter, placed in time about a month and a half later, means that almost without much effort on her part (somehow by sheer luck, or by unexpected and very favourable circumstances) things have taken a fortunate turn. That her whole year was neither easy nor relaxed is evidenced, among other things, by Saturn's position in the second house, the house of value, and perhaps also money (for example from sponsors, who were not in abundance in the middle of this year). Saturn is mainly important in this year's horoscope because it is placed as the first element after the Ascendant (not including the natal elements).

The two Venuses in this horoscope are activated at about the same time, positioned in the first half of August 1959. It was others (natal Venus in the first house and solar Venus in the eleventh house, the house of friends) who were assuring her she was on the right track and would succeed, rather than herself.

Knowing that the elections were successful, having taken place just before her birthday, it can be assumed that the beginning of her next solar horoscope would show us the outcome. And indeed, in the next solar return chart for 1959/60, Jupiter is placed at the very beginning of the year, and in the very place of great success – in the tenth house, the house of career and achievements! So, on her birthday, she was overwhelmed by a feeling of success as a result of an event or activity, unfolding just before her birthday.

SOLAR RETURNS IN PRACTICE: MARGARET THATCHER'S BATTLES AND VICTORIES

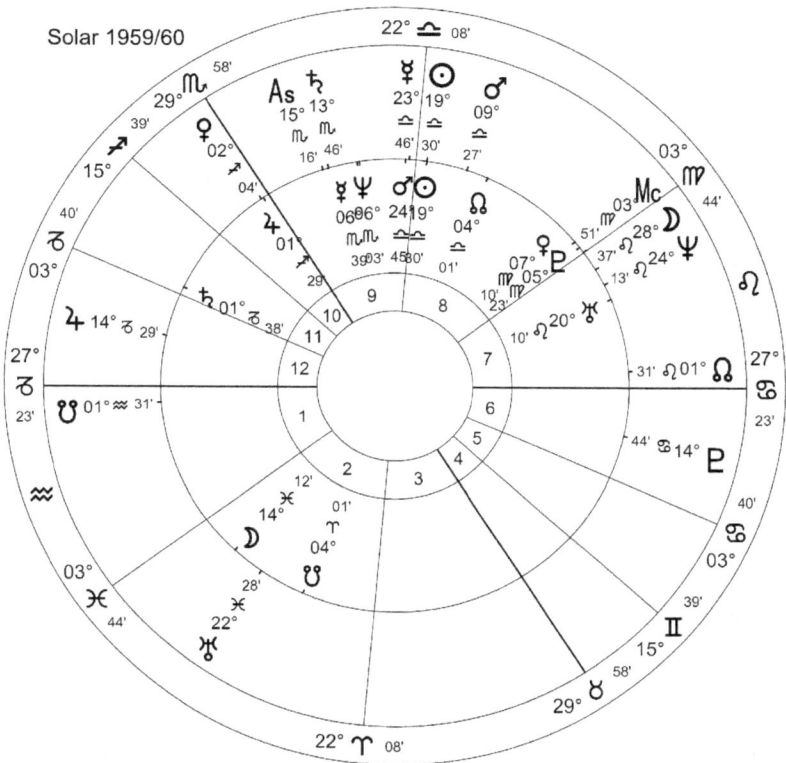

However, at the same time, other houses of the solar return chart are also highlighted, especially the eighth and ninth, and a little later, with the help of the Lunar Nodes, the first and the seventh. In the ninth house we thus first find natal Mercury, ruler of the 10th, 11th and 8th houses, and next to it, only a few days later, Mars, which, in this year's chart is not the Ascendant ruler, but rules the MC and the tenth house! In the 10th house, just after Jupiter, we find natal Venus, ruler of the natal 7th and 12th houses (support, trust of others, commitment to understand and work for their interests), and in the 8th house Pluto, ruler of the MC, and Venus, ruler of the 4th and 9th houses, are placed at the beginning.

And family? Margaret had two small children after all... They couldn't have had the best of times – the ruler of the fifth house here is Mercury, closely conjunct Neptune high in the ninth house and in Scorpio, not exactly a family idyll. In addition, almost all the planets in this solar return are found in the upper, ambitious half of the horoscope, indicating success and upward mobility. That alone counted.

And Denis? The seventh house cusp is in Cancer, embracing the whole of Leo, so we have to check both the rulers. The Moon is in Pisces and in the second house - Denis was certainly supportive of her, but there were big expenses (the eighth and second houses are indicative of expenses or talk about family money), besides the nanny Abby having been spending more time with children than their own mother... The Sun, co-ruler of the seventh house, is placed in the eighth house, certainly indicating a marital crisis, which can be located in the autumn of 1960; however, in the seventh house itself it is announced even earlier, through Uranus (in the first half of June), as well as natal Neptune and the Moon, which activated the later months. It is true that the natal North Node shows that Denis had been aware of the true nature of his marriage already by November 1959, having been unprepared for, or unable to cope with, such profound and, above all, rapid changes in his life.

This is confirmed by the events, taking place a few years later, when, just after the 1964 general election (in mid-October), Denis suffered a nervous breakdown "due to loneliness and exhaustion" and went to South Africa for two weeks to recuperate. In the 1964/65 annual horoscope a stagnation rather than progress in Margaret's career is visible (emphasis on the lower side of the horoscope, with Saturn as a brake in the 10th house). In addition, Margaret was now (had to be) investing her energies in the fields, where she had not been so involved previously, the home (Mars in the 4th house) and the children (natal Mars in the 5th house, both within a few months of the birth date).

The victory

The most crucial year in Margaret Thatcher's life was 1974/75, the year she won both the party election (December 1974, or the second round on 11 February 1975) and the general election a month and a half later, on 28 March 1975.

In this horoscope, which is very illustrative in terms of these key "Iron Lady" upheavals in life, let us first examine the first of these dates, December 1974. It is a well known fact that Mrs Thatcher won the key points in her programme at the expense of the criticism aimed at the then Prime Minister, Edward Heath. However, a look at the previous solar return, the one for 1973/74, when she was preparing for the campaign, it shows Mars, one of the key elements in her birth horoscope, placed together with the Moon in Taurus in the tenth house, as a symbol of combativeness. In this horoscope, however, there is still a major focus on the fourth house, denoting home, family and children, but also traditional values and a sense of patriotism. With a Leo Ascendant Margaret must have been very successful, strong and unstoppable this year, however, unfavourably aspected Saturn in the second half of the 11th house threatens to annihilate some dreams, heart's desires, plans. Its aspects, a square with Pluto in the third house, and an inconjunction with Venus and Neptune in the fourth house, have obviously brought the necessary challenge for a most stubborn struggle; for it should not be overlooked that Saturn also has an excellent sextile with Mars in the tenth house, signifying, perhaps, her determination to win at all costs.

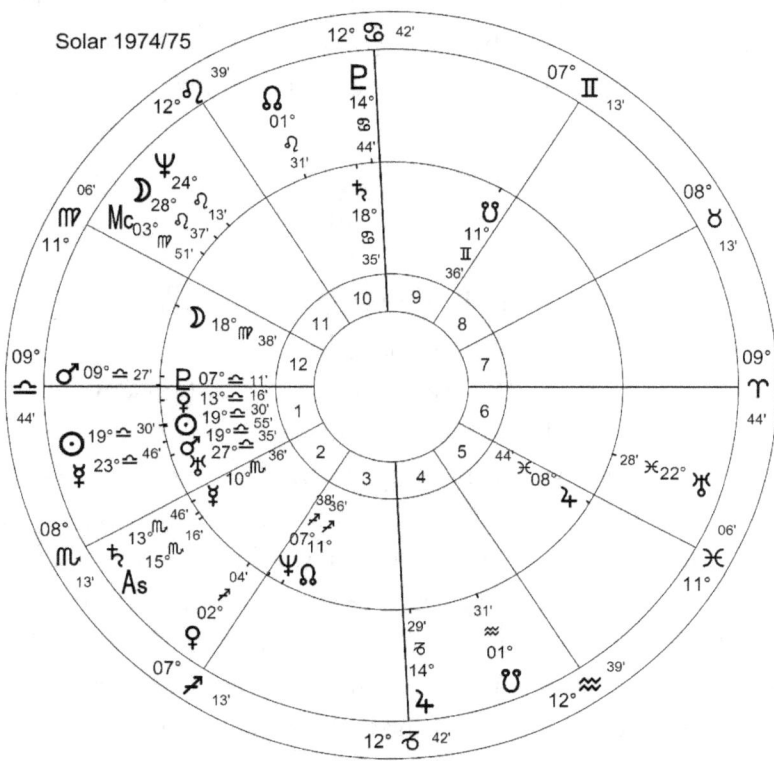

In the next, 1974/75 solar return chart, we should first of all have a look at December, the time when she tried to take over the leadership of the party, but then, as we know, only succeeded in the second round. If there were an obstacle, a brake, in the way of her success here, it is exactly where expected - Saturn in the tenth house! Counting the degrees and converting them into months results in December 1974! She did not make it, not yet. Saturn decided the time was not right yet. And being the god of time, he will let her go ahead when he so decides.

He did, however, let her go ahead in the second round, in February the following year. Yet this solar return lacks any indicators of victory whatsoever in the tenth house. Instead, a half-degree exact conjunction of Sun and Mars reigns in the first house at exactly the same time, with Mars in the first house signifying the battle, the victory, the general. And since the Sun in the natal horoscope is part of the stellium, which is also the apex of the T-square, all three are placed right here in the first house at this time. Natal Mars is even exactly on the Ascendant, adding to activation, and the Sun and Mercury together only a little - a few months – later, coinciding with the time of victory.

But if the former Margaret's pivotal years had Scorpio Ascendants, this time it is Libra. Unexpected? Libra in a yearly Ascendant, after all, means "a year of harmonisation, balancing, appeasement", but here we are dealing with the victory of a lifetime?! How is this possible? It is easy to imagine that in the mind of the "Iron Lady" - after so many years of sacrifice, effort and hard struggle - the feeling of "I am finally where I belong" or perhaps "everything just fell into place as it should have fallen long ago" almost certainly prevailed after the victory. Triumph often means levelling out.

And the third term, winning in Parliament? Here, too, the indicators are at least a little surprising. If this term falls almost exactly on half of Margaret Thatcher's solar year, almost nothing can be found in the half houses of this year's horoscope - with the exception of **natal** Mercury, which activates the first house just then, as well as the natal Moon, which occupies the middle of the eleventh house of future vision, and which in the natal horoscope is culminating close to the MC from the ninth house. This is the solar return of her destiny finally coming true, bringing – victory and glory.

The message of this year's horoscope, then, is that the "Iron Lady" Margaret Thatcher felt that she was finally, where she was meant to be – at the top. And at home? The concentration of virtually all the natal and solar planets on the left half of the horoscope shows that everything was geared to her success this year, and that she managed to keep everything under control, so no wonder that home and family seemed to be rather out of favour. The fourth house in Capricorn, the fifth in Aquarius, the sixth in Pisces and the seventh in Aries – all these together do not suggest these spheres of life to have been at the forefront at the time. Her husband had changed a lot in the twenty-five years of their marriage, having been very supportive and co-operative, while the children had gone their own ways - Aquarius on the fifth house cusp and Uranus, the ruler of that house, in the first house (she did care, though...). What is more, natal Uranus, in the birth horoscope in the fifth house, the house of children, is now in the sixth house, meaning a sudden change of home. Since it is a natal planet placed in the outer circle, it is an outside person, who has caused this.

It is known that at the time, Margaret's daughter was preparing for a major move to a distant place and actually left the following year. Natal Saturn, ruler of the third house, placed in the second house here, makes it clear that someone from outside was causing her problems (for example financial), depreciating her value at the very time when so much was happening in the political sphere. Perhaps one of the children's lifestyles has come at a considerable and unwanted cost, for example her son's involvement in car racing...

Configurations

In solar returns special focus should also be given to interplanetary configurations, which combine a lot of planetary energy and also show, which fields are interconnected. This in turn means that it is often possible to understand, why a person perceives an event in one way and not in another, perhaps "normal" or even expected one. And yet, an event taking place in May, may have been conditioned or even triggered by another event back in February, and may even result in an event occurring in November - if only the planets that mark these events are interconnected by aspects. Thus, individual parts of the configuration may "fall" into different periods of a solar return.

It is also important to remember that a configuration within a solar return itself is not very significant, unless also linked to at least one significant natal point, luminary, planet, Ascendant, Midheaven or Node, and vice versa – a natal configuration is with us throughout our lives, so it can be observed in every solar return. Explaining the significance of a (natal) trine in a particular solar return is therefore of little use, unless the trine is related to the solar planets or points. However, if there is such a connection, the configuration suddenly takes on great significance.

There are many possible combinations. Perhaps a solar planet is placed on one of the points of a natal configuration, perhaps a whole configuration of planets activates a natal planet in a solar return, thus getting a prominent place in the interpretation of this year's horoscope, and perhaps a few planets from the natal chart and a few planets from the solar return chart make up a configuration, not appearing in either of the two per se.

In the present example, Margaret Thatcher's solar return horoscope for 1974/75, there are three such configurations:
 - solar Uranus is conjunct natal Mercury, also part of the natal *yod*, which it forms with Neptune and focal Uranus in the fifth house;
 - solar Pluto and Venus conjunct the solar Ascendant and Mars, placed right under the Ascendant, while Venus also conjuncts the Sun, which is the focus of the natal T-square, a key configuration in Margaret Thatcher's horoscope. This natal T-square is also acti-

vated in this SR by solar Saturn, placed in conjunction with Pluto in the tenth house;

- solar Mercury and Saturn, which were not in mutual aspect at the time of the birthday, each conjunct natal Saturn and Pluto, which are in aspect (trine), and Jupiter, placed in the fifth house, forms a Grand trine with three of these planets, amplified by the solar MC.

All things considered, this is a truly remarkable annual horoscope, which highlighted all the strong potentials of the natal chart and even added new ones that enabled Margaret Thatcher to rise to the top. Since, generally, great success requires a combination of both unfavourable as well as highly favourable planetary energies, the *yod* and especially the T-square brought plenty of tension and challenge, i.e. the basis for a hard and relentless struggle, while the favourable (and strong) aspects of the Grand trine at the same time offered plenty of opportunities for a favourable outcome.

Prime minister

According to the Chapter 2 table, the Sun enters a sign at about the same time after four years, which means that annual Ascendant is also somewhere around the same degree after four years. This in turn means that a solar return after four years is to a certain extent similar in qualities and orientations to the previous one. This is also evident in the case of Margaret Thatcher, the solar return for 1978/79 again having a Libra Ascendant, and the whole chart having a strong emphasis on the left side, i.e. on the sphere of "Me". Natal Mars and Sun are again in the first house with natal Pluto in the tenth, all together indicators of power and success. This horoscope proved very useful for the year when the 'Iron Lady' became the first female Prime Minister in the history of Great Britain.

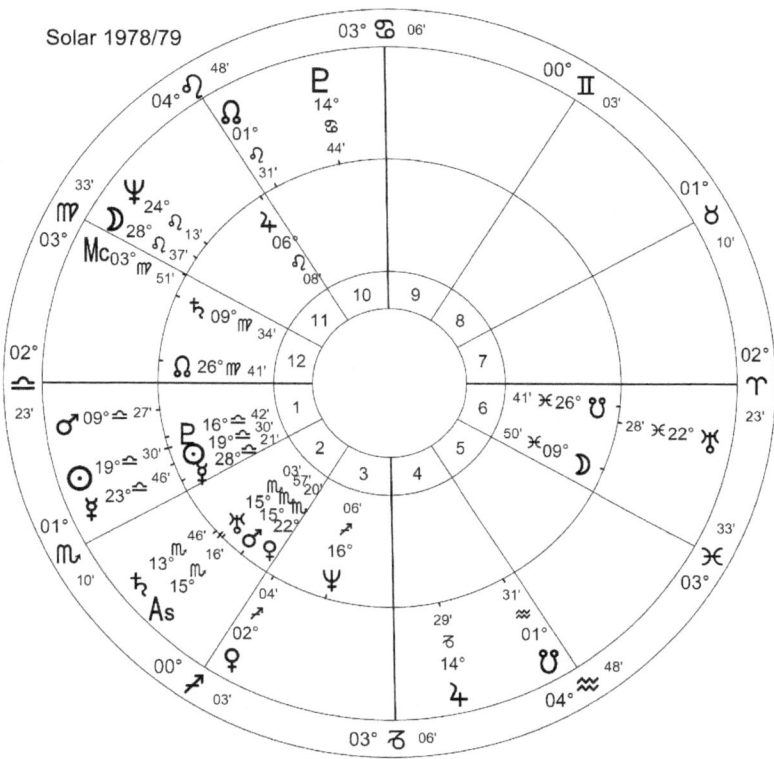

This occurred on 28 March 1979, almost halfway through the solar year. At that calculated time, Pluto in the first house and a tight cluster of four elements, solar Uranus and Mars, along with natal Saturn and the Ascendant in the middle of the second house, are placed in this horoscope. Natal Pluto in this chart points to a time somewhat before this victory, and can be interpreted as a 'strong obstacle from the side or from the above' on the outside of the tenth house, which, with the otherwise great propulsive mode of this horoscope, could not have been strong enough to make Margaret surrender in this struggle...

Victory in war

Matching the importance of the so far examined years as regards Mrs Thatcher's career is the year 1982, known as the year of the Falklands crisis and war. And also the 1981/82 solar chart return is, at first sight, a very special astrological chart. The first thing to notice is the focus on the lower part of the horoscope, which usually indicates 'introversion, passivity, regression, preoccupation with retrospective matters, or with family, parental or possessive themes'. But not in this case! This is the Prime Minister of a large and powerful country, so in such a person's horoscope the fourth house (extremely strongly represented here) and the Moon will signify patriotism, the country and its interests, rather than one's emotional world and home! In the fourth house there are as many as five planets ruling as many as eight houses in this year's horoscope, along with two intercepted signs. Of particular note is the close conjunction of the Sun, giving this house even more emphasis, and Jupiter, ruler of the tenth house of career, representing the key point of this chart. Sun and Jupiter are temporally placed sometime before the middle of the year, which in Margaret's case would mean the end of February and the first half of March 1982.

At the same time, another strong focus in this chart is placed on Mars in the second house as the leading planet and as ruler of the intercepted Aries in the tenth house. It is meaningfully conjunct the Moon, culminating in this horoscope as the chart ruler (again the homeland, patriotism), placed in the tenth house and in feisty Aries. Together these planets signify combativeness and determination to win, being a true indication of a leader, a warrior, a general. The position of the two Nodes right on the horizontal axis is also interesting, denoting the "meeting with one's own self". Knowing that the war started on 2 April, there can be no doubt about the perception of the situation, shown by these planets.

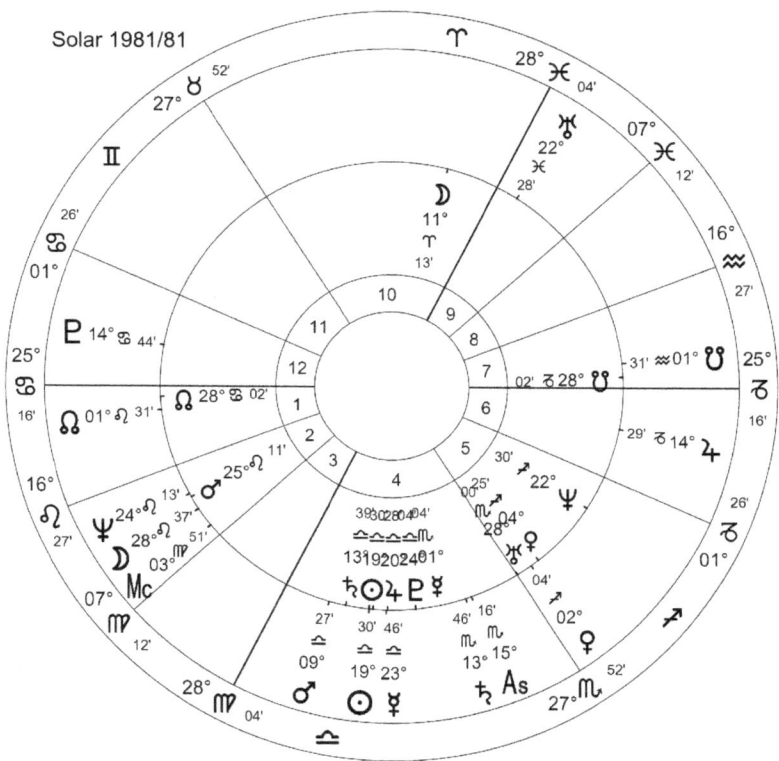

Further in the year less hard planets are found in the respective houses: the natal Moon in the second house, just after Mars, signifying recognition for attitude and victory (the Moon here is placed on the outer circle, meaning others), and Mercury, placed in the fourth house as the last planet of the five, which rules the third (communication, news) and fourth houses (homeland), and co-rules the eleventh house (future, development of the situation). Both planets are placed at a time which can be defined as "June 1982", the time of the end of this war.

The victory in the Falklands War was widely seen as decisive for the new parliamentary victory of the "Iron Lady" the following year.

Solar return 1982/83 has most of the planets on the left, especially in the first and second houses, indicating great confidence, strength and stability.

Some more important annual horoscopes

In 1984, one day before her birthday, the Irish Revolutionary Army planted a bomb in a hotel (in Brighton), where the Conservative Congress was being held, which exploded when the First Lady was not around. It is well known that the 'Iron Lady' demanded that the Congress go ahead as if nothing had happened - very fitting for her nickname. For the solar return timing theory, this event is a perfect illustration of a "birthday event not showing up in the solar return chart", perhaps in part also due to her intransigence. In fact, neither in the 1983/84 solar return nor in the following year's horoscope do the very house cusps show any strongly highlighted elements that could be associated with assassination or political terror.

Among the solar returns that remain to be mentioned, let us have a look at the year in which Margaret Thatcher resigned from power. This was in November 1990, when her popularity began to plummet, and she was in danger of losing her position even within her own party. The Moon is placed at the beginning – only a month after her birthday – of the 12th house (powerlessness, depression), the position of the MC ruler, Mercury, in the natal horoscope. Natal Mars in this return chart is in the second house (as co-ruler of the Ascendant: "what, if any, is my value?"), and natal Pluto, the other co-ruler of the Ascendant, is in the 11th house - "what does my future even look like?"

The solar return of 2002/03 is also very telling, because on 29 June 2003 her husband Denis died after a long illness. They had been married for fifty-two years and she often said he meant a lot to her and that without him she would not have been so successful. In her natal horoscope, Venus rules the seventh house and is placed in the first house as the leading planet - proving these declarations of love to be true.

In this solar return, the corresponding planetary positions in either the inner or the outer circle would be expected to be somewhere in the three quarters of the houses' width, corresponding to the difference between 26 June and 13 October. In the second house (being the eighth house of the husband), Mars, ruler of the tenth house (status) and also of the husband's fourth house (house of the end), is in the expected position, indicating the loss of partner. In the first house, exactly where Venus is placed in the natal horoscope as the ruling planet and also ruler of the seventh house, is the natal MC - an indicator of status or a change in status related to the husband. In the eleventh house natal Pluto is in the corresponding place as ruler of the natal chart, denoting her perception of the lost future.

There are also other illustrative positions, testifying to the perceptions and feelings of Mrs Thatcher after the loss. Pluto 'reigns' in the house of home ('the home destroyed'); towards the end of the 10th house (status), in the part ruled by Mercury, ruler of change, is the North Node ('it was meant to be'; a change we did not want); there is Mercury as the ruler of the 11th house, which contains Saturn, and the Moon in the 5th house, and Jupiter in the 12th house, here the ruler of the 5th house - 'children who have been my main support in these moments'. An essential element of the chart, the Sun, ruler of the Ascendant, is placed in the third house – a major change in attitude or immediate surroundings (or both) is the key focus of the year!

* * *

Finally, it makes sense to mention once again the rule applying to astrological prediction in general, namely, that to confirm our findings, other techniques, such as transits or progressions, would be useful; however, this is not the purpose of our book, as the text would then become too long and complex, as well as less clear. In any case, this rule is valid and should never be overlooked in prediction, solar return being just one amongst the established predictive methods in astrology.

X

CASE STUDY: WHATEVER HAPPENED TO PAUL

As another example of the use of solar returns, let us examine an event that may not even have happened. At least according to the official version, although some authors have been trying to prove for decades that Paul McCartney, as we know him today, is an imposter.

According to many, the famous Beatle was fatally injured at a London intersection in the autumn of 1966, during the recording of the *Sgt. Peppers Lonely Hearts Club Band* album. There are many stories, proofs and "evidence", comparisons, studies, theses, claims and much more on the web. A definitive answer to this question is unlikely, not least because the actors involved are increasingly fewer and fewer - two of the Beatles, John Lennon and George Harrison, are now deceased, as is their manager, Brian Epstein, who died of an overdose of sleeping pills in the late summer of 1967.

The applicability advantage of solar returns and the simplicity of the timing method can also prove useful in uncovering a veil of mystery in a complex case such as this, as it can help us try reconstruct, what had actually happened. Namely, if we can rely on the thesis – thousands of examples confirming this premise – that a solar return also shows the mood of a moment, or rather, our perception of an event, although not the quality of the event as such, then, perhaps, through the available data of the actors involved in the story at the time, we can form an opinion as to the possibility or probability of the event, without, however, getting a confirmation that it had really taken place.

THE FATAL RED LIGHT OF A NOVEMBER MORNING

Let us first examine the situation. All the accounts of this concealed death are based on the fact that Paul McCartney was killed at exactly 4.38 am on 9 November 1966, while riding a car on the streets of London after a major quarrel with the other band members during the recording of an album. Under the influence of LSD, in a company of a random friend, he allegedly ran a red light and crashed into a telephone pole after colliding with another vehicle. Within four minutes of the crash - in which his completely accidental passenger, teenager Carolyn Rita Northam, was reportedly killed instantly breaking the windscreen with her head - the car burnt down and passers-by were unable to help the knocked-out and fatally injured Paul. He was pronounced dead at 4.56.

CASE STUDY: WHATEVER HAPPENED TO PAUL

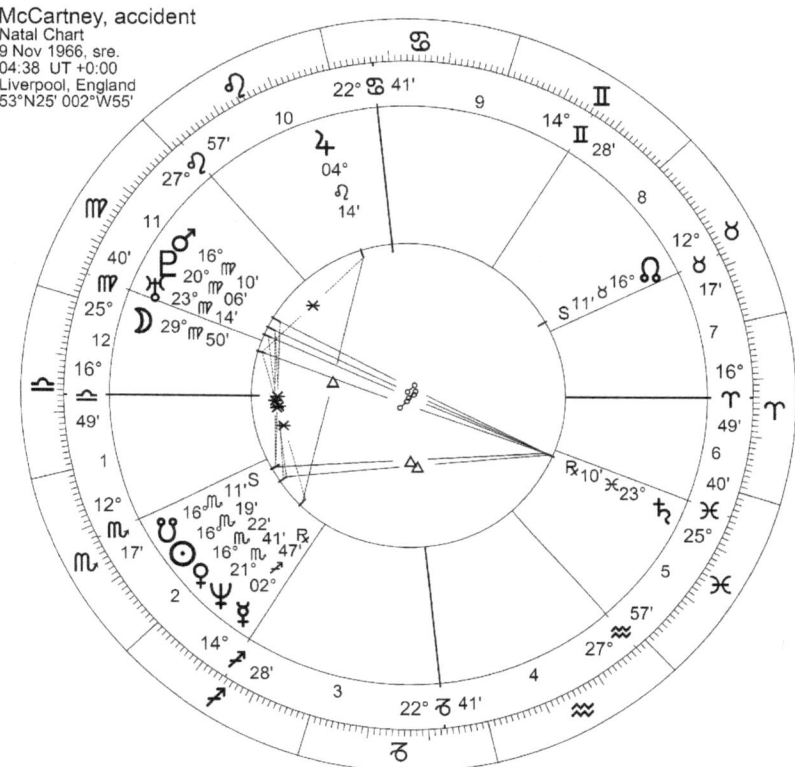

The astrological chart of the accident itself is quite illustrative: it is hard to say which is more noticeable, the opposition of a lone Saturn with a wide stellium of Mars, Pluto, Uranus and the Moon, all in Virgo, or the extremely tight stellium of the South Node, Sun and Venus (within 0°11'!) in Scorpio, simultaneously aspecting the opposition. At a first glance at least, this chart is not a "typical astrological chart of a car accident", being enclosed only for relating to other charts.

The next step deals with our hypotheses, based on two basic premises: 'Paul had an accident' and 'this is all a hoax or sensation-mongering by sensation-hungry individuals and media'.

How would other Beatles have perceived the death of their colleague? Given that they had been working together and had big plans just an hour or so ago, it would certainly have come as a major shock. It is known that their manager Epstein broke the news to them about an hour later, in the early hours of the morning. After all, these were very young guys in their mid-twenties. A friend, a colleague would have died, together with their future - McCartney having authored, along with Lennon, the vast majority of their music. In their horoscopes one would expect the influence of Uranus (suddenness, reversal, shock), Pluto (death, destruction), Mars (blood, violence, death), Nodes (fate), Neptune (drugs), perhaps Mercury (traffic), in different unfavourable combinations - squares, oppositions, inconjunctions. The death of a loved one is often very strongly marked in our horoscope, depending on the manner, the relationship, the degree of expectation and more. And Paul was **undoubtedly** a close person.

And if it were really a hoax? Uranus and Mars would not be indispensable here, Neptune and Mercury (bluff, lie, deceit), ill-placed Venus (benefit), perhaps Pluto (feeling pressured by an injustice) playing a more prominent role in such horoscopes; the rumours of a hoax, however, had only emerged afterwards, so there could not have been any major reflection of deception in the horoscopes **at the time the accident was supposed to have happened!**

Astrologers are therefore only left to examine the charts of the respective date.

FIVE MORE HOROSCOPES

As always, the charts of all the five participants are relevant, that is, the members of the band and the manager Brian Epstein.

Paul McCartney could briefly be described as "very ambitious, famous at an early age, desirable, self-confident, eager to be affirmed and loved, with great artistic talents, the finest of which having been music". Could a person with this horoscope die a violent death or in a car accident? Very easily! The rulers of the third (traffic) and eighth houses (death) are closely conjunct in the 11th house (suddenness), and the other two Malefics of the horoscope, Saturn and Uranus, are also closely together in Gemini (again, a combination of communication, traffic, and suddenness). Jupiter, placed high, conjunct the Sun (self), squares Neptune on the very Ascendant – a shortcut to drugs and mistakes.

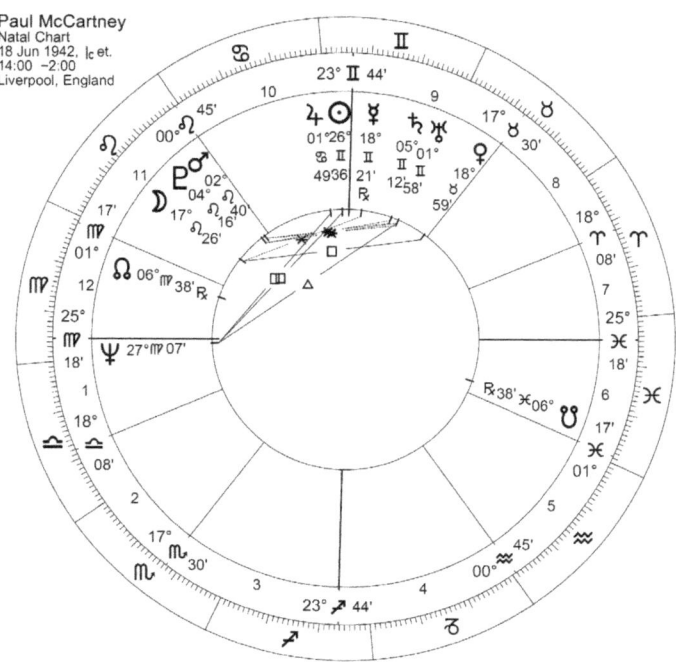

In **John Lennon's** chart (Libra with the Moon in Aquarius and also in the 11th house, denoting futuristic mind) with Aries ascendant, Mercury in Scorpio as the focus of the T-square with the Moon and Pluto is pronounced – a touch of mystery? Mercury so placed never loses its head, even in the most difficult situations. And it was Lennon, who (according to sources) suggested the affair to be covered up. Could a righteous Libra react like that? With unwanted events, Libra's primal impulse would always be "no, that didn't happen/that should not have happened". But Lennon had several oppositions in his horoscope, including the combination of mutually opposing Libra and Aries, with (quick thinking, different, detached) Aquarius, offering itself as the ideal solution here - "let's do it differently" and "let's do it now!"

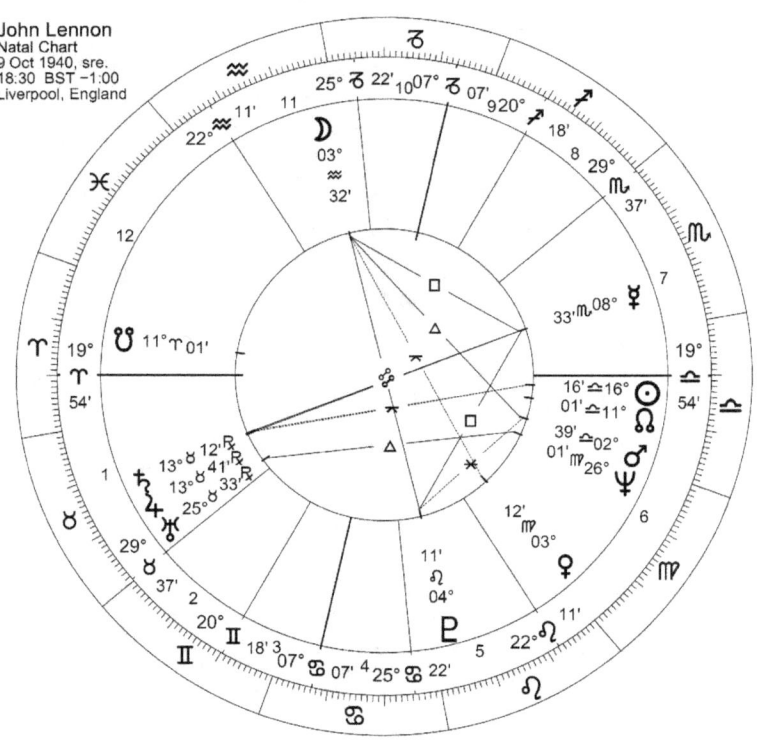

CASE STUDY: WHATEVER HAPPENED TO PAUL

Ringo Starr's chart contains all the ten planets in the lower half of the horoscope, which, together with the Sun in Cancer and the Ascendant in Pisces, give a picture of a man, keeping in the background and leaving leadership and fame to others, with a few other indicators pointing to success later in life.

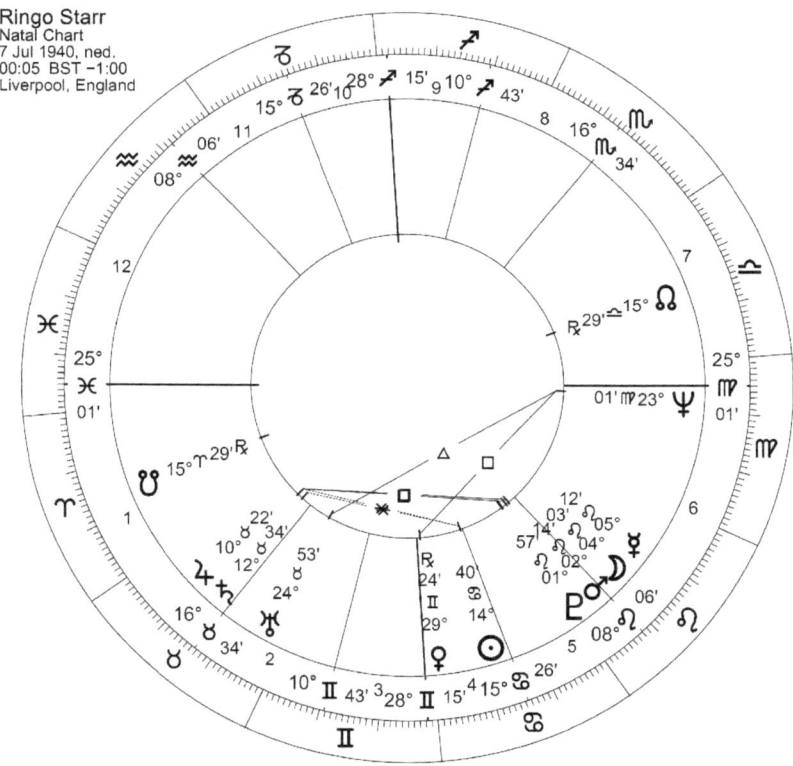

The chart of **George Harrison**, the youngest Beatle, also shows a predominant combination of Pisces and Cancer, extreme subtlety and sensitivity, swinging from one extreme to the other, many talents, but not a great sense of achievement in life.

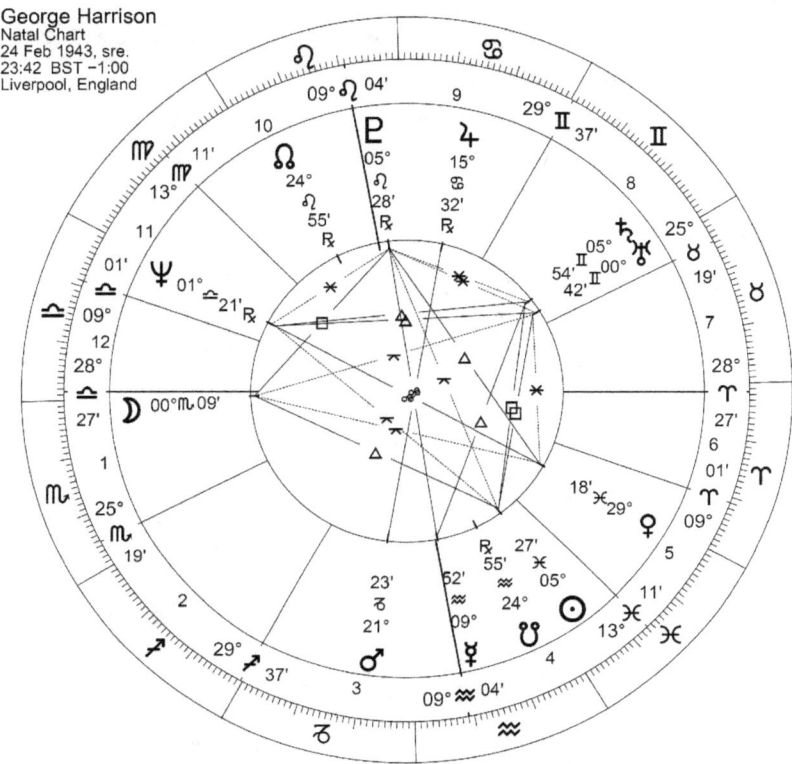

But there is a fifth member, the manager. The French website *astrotheme.com* has his horoscope in its vast collection of horoscopes, unfortunately without quoting the source. His is a truly special natal chart if reliable. A Virgo with a Libra Ascendant and a Grand Cross, a configuration that often obscures in meaning and power the basic principles of Sun, Moon, Ascendant. A Grand Cross of this quality is indicative of inner strength, charisma, firm intentions, as well as, to some extent at least, of dogmatism, determination and unwavering convictions. The configuration consists of Jupiter, Moon, Uranus and Pluto. **Brian Epstein** was strongly focused on success and "higher goals". He is even supposed to have participated in the conspiracy,

CASE STUDY: WHATEVER HAPPENED TO PAUL

or even helped convince the sceptic members of the group that its end could imply very heavy legal and financial consequences.

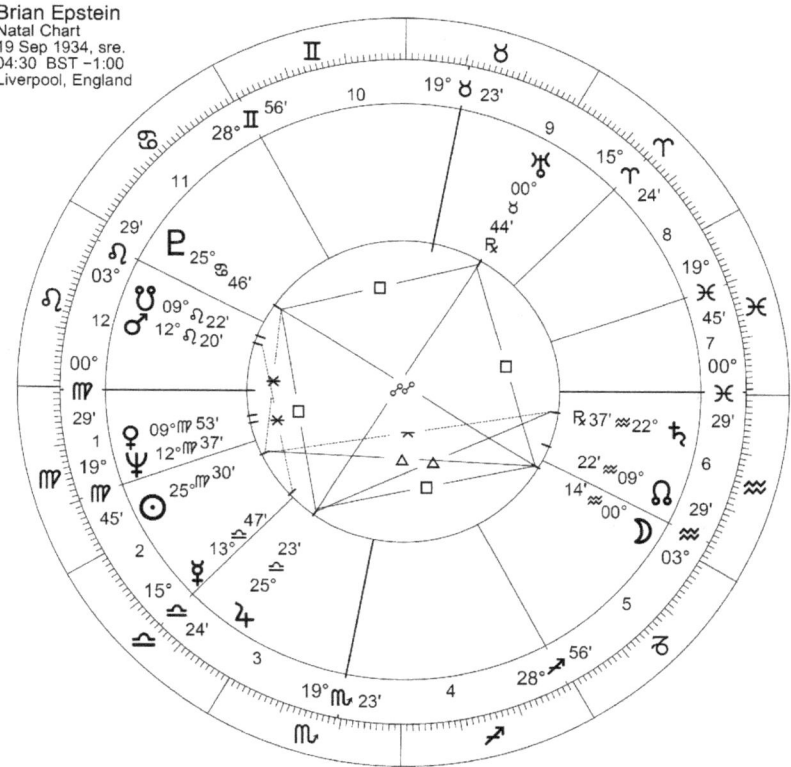

WHAT DO THE TRANSITS SHOW...

There are several ways for astrologers to correlate the charts of the participants with the chart of the accident itself, of which they had been duly informed. Usually they would use transits, progressions and directions. To repeat a well-known rule of prediction, a situation must first be confirmed by more than one predictive technique, before a prediction can be given with any certainty whatsoever.

However, since this would take up far too much space, let us only examine the transiting planets.

In Paul's horoscope, the Saturn-Neptune trine from the car accident horoscope directly activates the natal Ascendant-Descendant axis, the Moon of this horoscope being in close proximity to his Neptune in the first house, and more specifically activating his Sun and Jupiter close to MC. Uranus (misfortune) is right at the Ascendant, Jupiter from the accident chart could hardly be more closely conjunct Pluto (2 minutes), and the tight conjunction of the South Node, Sun and Venus is at the very end of the second house, opposing the North Node (fate) at the very end of the eighth house (crisis, death). But let us not forget: since we are talking about transits, the Nodes are mostly retrograde, meaning that the North Node is only well **into** the eighth house, not preparing to leave it. And there it immediately opposes the Sun and Venus at practically the same point, creating the precondition for a tragic event.

In John's horoscope, Jupiter activates Pluto (ruler of his natal 8th house) within three minutes of exactitude (!), along with the entire T-square, including the cold-blooded Mercury! The tight Scorpio stellium activates his Saturn (ruling the 10th and 11th houses, Lennon having the Moon in the 11th house), Jupiter in the first house, and through the 30° semi sextile also the Sun (at the same degree of Libra, a partile aspect).

George's horoscope has relatively little crossover with the horoscope of the accident. The Moon and Mercury activate Neptune in the 11th house (friends). Apart from that, Neptune (disbelief?) squares George's Node, also attacked by Saturn via inconjunction. Did Harrison's subconscious mind only perceive and accept the whole thing later?

Ringo Starr's horoscope, however, was highly activated by this event. Saturn of the accident chart activates the natal Ascendant (from the 12th house), the Lunar Node activates his Saturn, Jupiter conjuncts the stellium of the four planets Pluto, Mars, Moon and Mercury, while Uranus and Saturn activate Neptune (ruler of the Ascendant), and the stellium in the horoscope of the accident touches Ringo's eighth house!

And Epstein's horoscope? Strongly activated by the planets in the disaster chart. Jupiter is placed exactly on the cusp of the 11th house (plans, future), with its involvement activating two of the arms of the Grand Cross (meaning "something must be done!"); the opposition of Saturn and Neptune directly activates Epstein's Sun in the 12th house (powerlessness, but also conspiracy and not least escape from an unpleasant situation), the Moon activates the Ascendant, and the tight stellium in Scorpio "attacks" through a square the manager's Mars, ruling money in his horoscope.

... AND THEN COME THE SOLAR RETURNS...

Paul

Paul McCartney's last solar return is not very likely to reflect the accident or the death itself; there was probably not enough time for such a thing [11]. We could, however, assess the yearly potential, as well as Paul's recent mood, to gain at least some insight into the chain of events, having led to the tragedy.

---

11   The question is what reflection can be expected in the chart of a man, who lived no more than four minutes from the time of his accident to the time of his death, and who was overwhelmed with fear and pain.

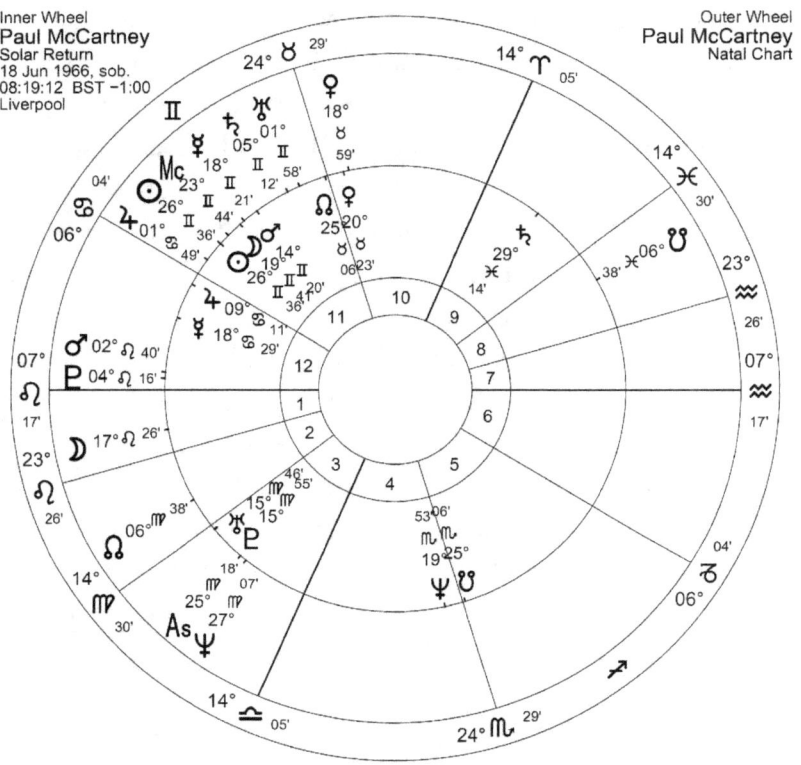

In general, this solar return chart points to a dramatic, unfavourable, perhaps even fatal year. Natal Pluto and Mars are close to the ascendant from the 12th house, and the four solar planets, namely Uranus and Pluto (in the 3rd house) plus Mars and Moon (in the 11th house) are in a violent square to each other; however that would have been insignificant, had they not activated one of the natal planets, in this case Mercury, natural ruler of communication and transport. In Paul's horoscope Mercury rules the Sun, one of the key elements of the horoscope!

If Paul's birthday was 18 June, then 9 November, the date of the accident, is just before halfway through the year. In this horoscope, Mars in the 11th house and in Gemini, in close square to Uranus and Pluto at the beginning of the third house, coincides with this time. Uranus, ruler of the eighth house (!), in the third house points to a sudden event in communication or transport, with the presence of Pluto and the square with Mars (the most violent aspect in astrology) additionally confirming the chance of a fatal accident.

The natal Ascendant and Neptune can be placed at roughly the same time, with Neptune involved in a very unfavourable T-square with the Sun and Saturn. Recall: Paul has Neptune on the Ascendant, squaring the Sun in his natal horoscope, the square upgraded to a T-square by Saturn, ruler of the 6th, 7th and 8th houses in this chart. His chart - with Sun and Jupiter on the MC, both squaring Neptune on the Ascendant - could also be interpreted as a picture of a "spoilt brat, to whom all doors are open and everything comes easily", but these aspects of Neptune can also be associated with intoxicating substances.

Before the fatal time, not long after birthday, in July or August, are there any new plans, opportunities, something new, different, visible in the chart? Something that is about to cool off, perhaps, or become impossible, unfeasible, unworthy, limited? These being natal planets, was it caused by someone or something else - other members or circumstances?

Knowing that the band had their last concert on 29 August that year (a planned and agreed last concert!), this should have been reflected in Paul's horoscope at about the fifth (= a good two months after birthday) of the respective house. Natal Uranus and Saturn in the 11th house are perfectly in line with the activity, however, it can also be speculated that these developments were not quite to

Paul's liking, as they are placed in the outer circle - someone else making the decision?

Also very illustrative are the aforementioned Uranus and Pluto in close conjunction at the very beginning of the third house (relationships), with Uranus as the leading planet even, there being no solar planet from the Ascendant onwards. Uranus is subversive, Pluto destructive - seeing that Uranus rules the seventh and eighth houses (others, colleagues, partners; their worth, joint money, investments, debts, etc.), can we therefore infer that Paul had a major falling out and a break-up with someone very soon after his birthday? With his partner, his girlfriend - or with his partners, the musicians?

And more: the nodal axis together with Neptune (deception, fog, blindness, oversight) squares his natal Moon, and Saturn from the ninth house squares his Sun! This is certainly a very illustrative horoscope, containing some of the indicators one would expect to find in the horoscope of a person, bound to have a (fatal) car accident in the respective year.

John

9 November is exactly one month after John's birthday on the calendar - so about one-twelfth of the house or a point of close observation.

If something special was to happen at that time, it would be expected to show up in the horoscope, and since the death of a partner - and prospects! - is highly important, it can be expected to be reflected in the horoscope not just with one point, the activator, but with several.

CASE STUDY: WHATEVER HAPPENED TO PAUL

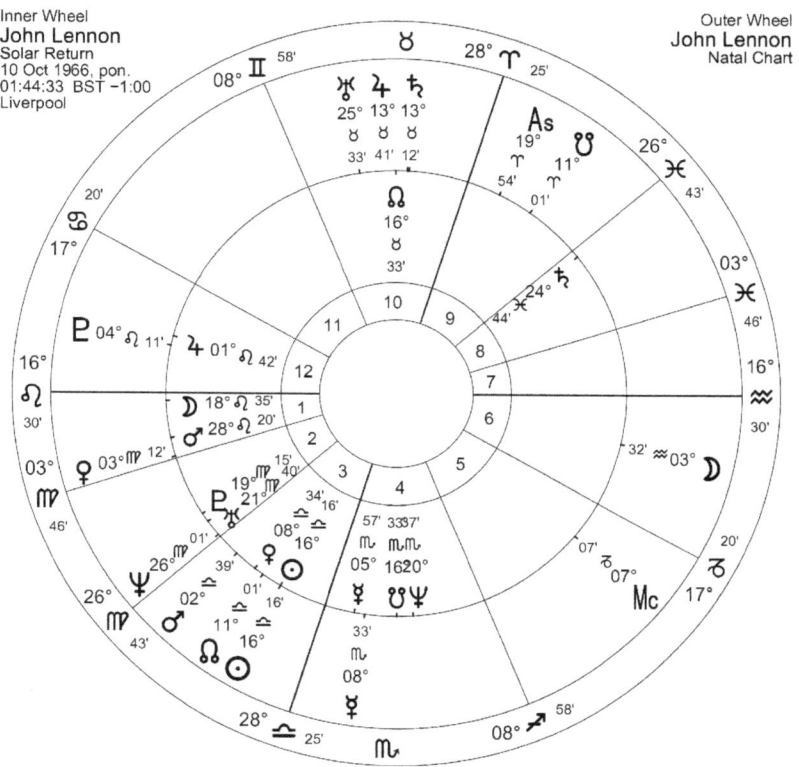

If we divide a house into approximate twelfths, in John's yearly horoscope only the Moon is in adequate position, quite close to the Ascendant, but squaring the natal Jupiter-Saturn duo in the tenth house, in the natal horoscope rulers of the houses from the ninth to the twelfth, and perhaps - depending on the accuracy of the birth time - even of the eighth house. The Moon, and even more precisely the Ascendant, square the rulers of practically the entire upper half of the horoscope! But John's tenth house in this horoscope is even more interesting. The two planets do not "activate" it until later, sometime in early spring, however, it first contains retrograde natal Saturn and Jupiter (perhaps a career standstill due to others, either people or circumstances?), then the North Node, placed sometime in March 1967 (perhaps "we know our path and we're moving on"), and lastly natal Uranus - a career turning point, but one due to someone else's decision, not his? But this horoscope is also illustrative in general, with a Leo Ascendant and the Sun in the third house, bringing many, major or dramatic changes.

George

With regard to George's birthday, 9 November corresponds to about three quarters of a year (more accurately, two weeks less), so the accident potentials should be sought at adequate points in the yearly horoscope. George's horoscope may be more telling than the previous ones, his birth time being accurate to the minute. Thus, the positions of the planets in his annual horoscope can be more precisely dated. At the time of the accident, the most accurately placed planet is natal Venus in the tenth house, ruler of the Ascendant, 12th and 8th houses of his natal chart, aspected by the Moon (in Scorpio; placed on the Ascendant itself, highlighting sensitivity and subtlety), by Uranus and Pluto, and also by Neptune, ruler of the Sun.

But there are a few other positions in the horoscope close to this date that may be strongly associated with the events, all of them natal, including Pluto in the second house, Neptune in the fourth, Mars in the seventh and Mercury in the eighth house. Temporally, Mercury is the first among them, followed by Mars - perhaps a realisation of the value of someone close and then a dispute about it (or perhaps about money and distribution, as the exact natal MC position in the second house of SR at the same time might suggest)? According to one of the several theses this angry departure of Paul from the studio and the subsequent accident may not have been the result of just "one argument", but rather a culmination of activity going on for a long time, perhaps at least a month...

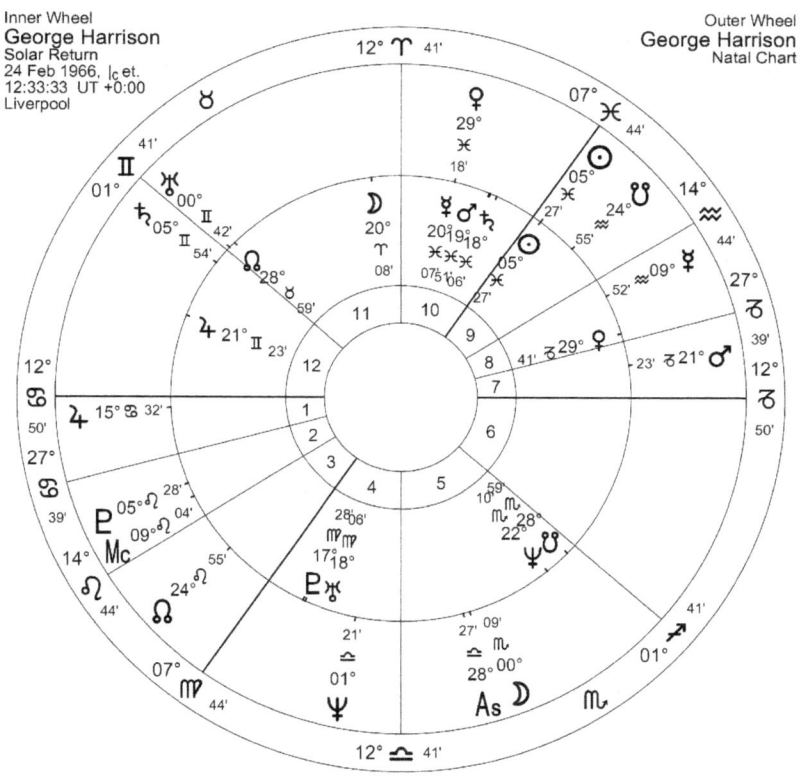

Ringo

In Ringo's case, the time between his birthday and 9 November falls practically exactly on 1/3 of the year, with natal Ascendant in the 9th house of the solar return accurately placed, as well as natal Neptune in the third house squaring Mars in the 11th house, ruling the 5th SR house. ("Neptune as a major threat") Otherwise, the 12th house is the most prominent house of this horoscope, containing the Sun, which, apart from rendering energy and tone to the chart, is also the Ascendant ruler. The ruler of the Ascendant in the 12th house can also relate to hiding, depression, pressures (the Sun and the stellium of the planets are in the outer circle, so imposed, "from the outside"), conspiracy? Does one have to learn to live with something to keep secret forever? (This is a demagogic statement, of course, but had such an event really have taken place, this would have been the expected response in the annual chart)

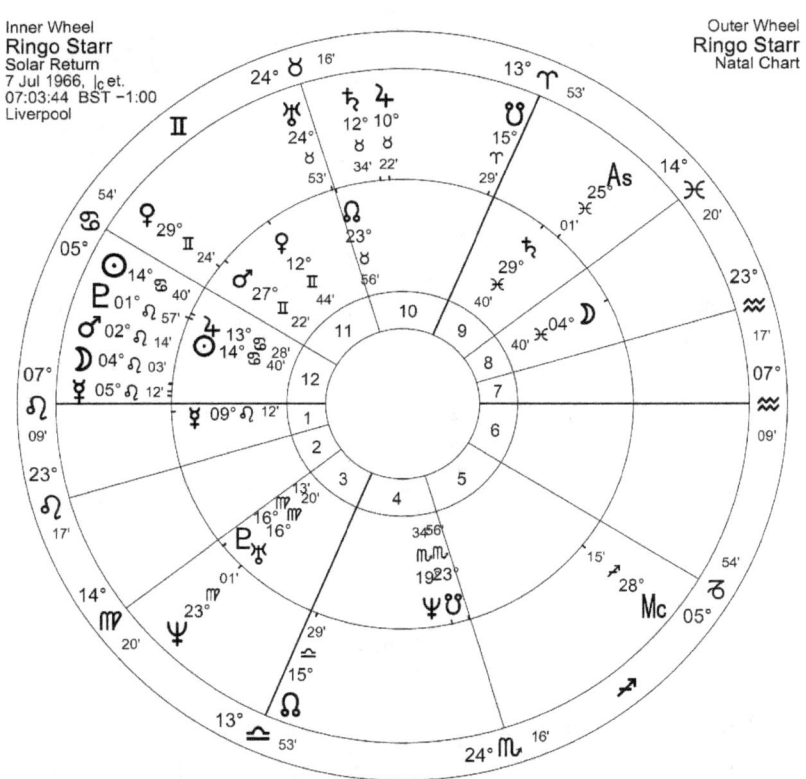

Brian

This solar return chart is also telling, especially if we know that it was the manager Epstein, who allegedly launched the idea and subsequently carried out a large part of the coverage of Paul's death.

First of all, both natal and solar Mars are placed at the cusp of the fourth house, which in Brian's birth horoscope is in Scorpio. This is all about survival! Besides, most of the energy is concentrated in the fifth house (stage, creativity, even project management), while the Ascendant is in Gemini, indicating changes in the coming year. The fifth house here is indicative of the dramatic activity.

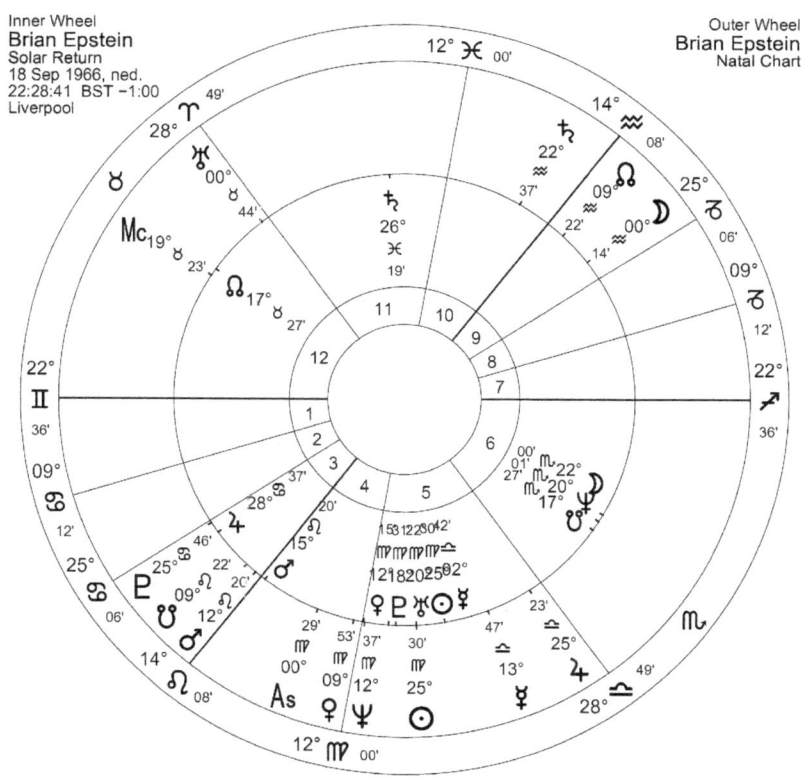

Very accurately with regard to the time of the accident, which in Brian's case is exactly 1/7th of a year, Pluto is placed in Virgo, squaring the Ascendant (and thus activating it), making a few other aspects, including a conjunction with Uranus. Uranus and the immediately following Sun in the fifth house both sextile ("exit") Jupiter in the third house and the natal Moon in the ninth house; according to the theories about this accident, Brian supposedly found a "replacement" for the deceased Paul abroad, in Canada...

According to the cover-up story appearing immediately after the accident, Epstein was extremely active making agreements, and, above all, bribing a lot to conceal the affair. Let us "search" for any highlighted activity in the solar return in the time following the accident date. Uranus, the Sun and later Mercury are found in the fifth house, natal Saturn in the tenth and solar Saturn in the eleventh, the North Node and natal MC in the twelfth house, the Moon and Neptune in the sixth - all up to the mid-points of these astrological houses, therefore until March 1967 ... certainly a very active horoscope!

Summary

There are many arguments against this incredible conspiracy theory, and as usually the case, there are so many convincing ones on both sides that it is difficult to opt 'for' one 'against' another. The most common objection is how a brand new man, a stranger, even a right-hander (Paul was a left-hander), could be taught to play, live, speak and everything else so well, that even people close to Paul never noticed anything strange. It may be impossible, however, after the accident the Beatles never had any concerts, where, with crowds and pressure, something would have been bound to be noticed - in fact, their last concert took place some two months **before** the alleged car accident; since then their work was completely studio bound.

And Linda, Paul's wife - she at least must have noticed! Not exactly, because according to the English version of Wikipedia, they only met for the first time on 15 May 1967, half a year after the alleged accident, meaning that Linda, an American, met William, a Canadian, posing as Paul. She could not have known the difference - at least at the time...

Since only two of the actors in this strange story are still alive, Ringo Starr and McCartney himself, they could clarify the matter themselves, but they are probably not interested. After all, the story could have fallen into oblivion after decades, had it not been for George Harrison, who, shortly before his death, decided to ease his soul and record an interview, relating his side of the truth. Was it a hoax, a deliberate deception? Unlikely, given that George had been referring to Paul as "Foul" in interviews for years, and in one recorded song the phrase "Paul is dead" is clearly audible... Had the matter been pressing George's mind for years, had it eventually contributed to his premature death from cancer? Maybe so, maybe not. Deceit and cover-up are activities depriving Pisces like Harrison, what with the Ascendant in the righteous Libra, of the sense of identity every Piscean has to struggle for hard.

And Epstein's death? He died nine months later, allegedly from an accidental overdose of sleeping pills. His solar return as a reflection of his perception at the time shows many influences and even more pressures, including the issue of homosexuality (Epstein would nowadays be called 'gay'), yet a possible explanation remains that the whole conspiracy could have been too much of a pressure for him too... The possibility of suicide is also clearly indicated in both his natal and his last yearly horoscope.

An astrological survey of these annual horoscopes shows a relatively high level of activity at the time of the accident, which would otherwise not have been logical. This is what makes one think there must be some truth in the story, which in fact seems quite probable. But until confirmed by someone with access to the data or a good memory, the truth will never be known for sure.

XI
WHERE TO CELEBRATE A BIRTHDAY

One of the most common dilemmas concerning solar returns is certainly the question of relocation, meaning "a different horoscope in a different place". Many people have moved away from their place of birth in their lifetime, sometimes literally to the other side of the world. The further away the place, the more chances to live in a completely different existential, informational and cultural environment, which in turn sooner or later affects our daily life, habits and character. A long-term and permanent relocation is bound to bring about personality changes, along with achievements that would probably not have been likely in our previous place of habitation. And when one returns home after many years, he is bound to be a new person, very different from his former self, or rather from the person having left many years ago.

A horoscope that shows these qualitative differences is called a "relocated horoscope", calculated for the actual location of habitation.

Compared to natal chart, it is - in accordance with the change of the basic parameters for a horoscope calculation, i.e. the latitude and longitude data - rotated in one of the directions, i.e. clockwise or counterclockwise, by a greater or lesser arc. If the Ascendant is chosen as the point of comparison, it will shift by approximately one sign with the latitude change of two time zones.

Before discussing a practical example, let us point out that a relocation of an astrological chart actually produces two "identical" charts, but with a different orientation. Planets in the relocated chart are in the same positions within signs as in natal chart, but they rule different houses and have different 'accidental dignities'. A solar Mars, for example, which in Pisces is not in a strong and good position, falls in the sixth house of the solar return, which is even less favourable; however, in a new, relocated horoscope, it is placed exactly on the Midheaven. This is certainly a strong position in the horoscope, and - associated with an angular point - is interpreted in a completely different way than in its previous position. Understandably, this Mars also contributes to a different quality of life.

Where to? Bali?

Let us examine the case of a person, who has lived and worked on the Indonesian island of Bali for more than a decade. After having travelled extensively around the world for educational and professional reasons, but without settling down anywhere, she decided at some point, and informed her family that she was going to move to Bali to start a new life. Comparison of the charts for the year before, made for both places, shows a practically identical

WHERE TO CELEBRATE A BIRTHDAY

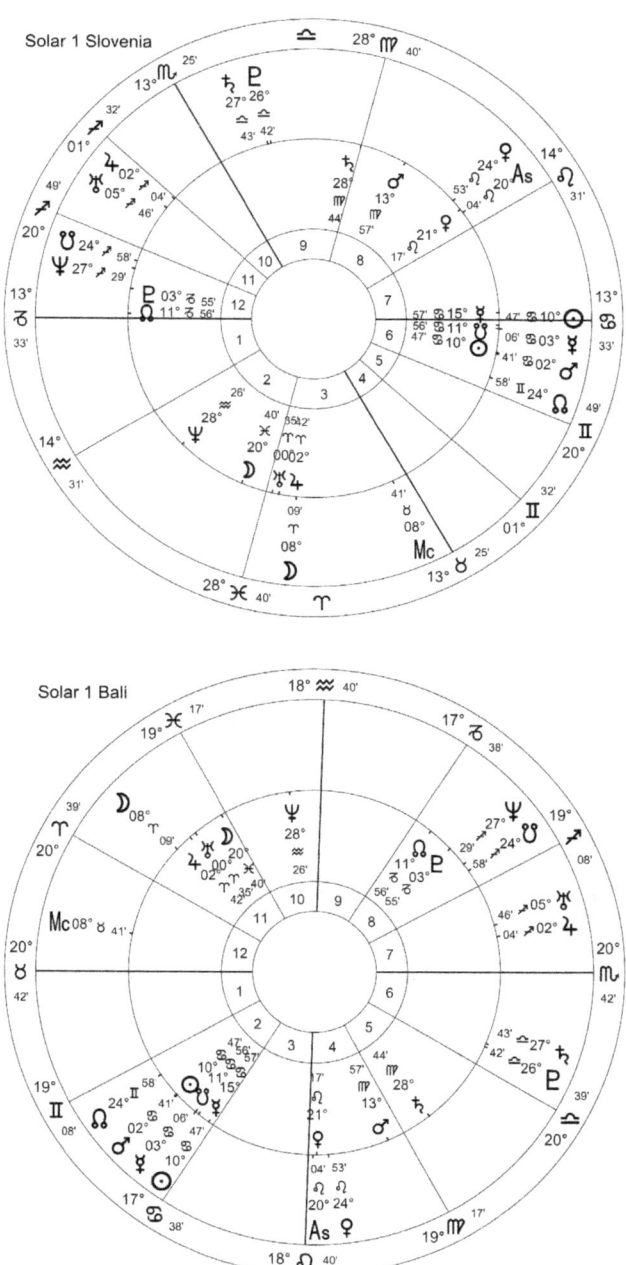

horoscope, except that it is rotated for a certain angle, the calculated Ascendant being different in the location with different longitude from ours, which also entails different positions of the house cusps. Thus planets remaining in the same positions within signs, and also forming exactly the same aspects as in the original chart, end up in different houses - and therefore also act differently.

Specifically, in the first chart there seems to be a distinct focus on Pluto in the 12th house, involved in a number of unfavourable aspects with other planets, a principle bringing severe pressure or powerlessness, as discussed in the previous chapter. Pluto in this chart opposes the Sun within a wide orb, but is in exact opposition to natal Mercury (pressure for change) and Mars, the natal ruler of the 4th and 9th houses, thus combining the concepts of domestic and foreign; both planets are also part of a wide stellium together with the Sun. Meanwhile, in the chart for Denpasar - the second chart - on the island of Bali, the 2nd and 11th houses are highlighted, denoting stability, money, future and vision.

The chart for the location of Ljubljana, meanwhile, contains another highly important potential, discussed earlier in the book – the so called "repeated position", rarely ever as exact as here. Let us have a look at the Uranus-Jupiter pair at the very cusp of the solar 3rd house (change), which in turn, as a repeated position (in the same sequence and almost equidistant from each other) in this year's horoscope, falls at the very beginning of the 11th house - diligence, plans, vision! And not only that: the two pairs are even in exact trine, offering Bali as a perfect "escape" from the hardships and traumas the person experiences in her home environment!

WHERE TO CELEBRATE A BIRTHDAY

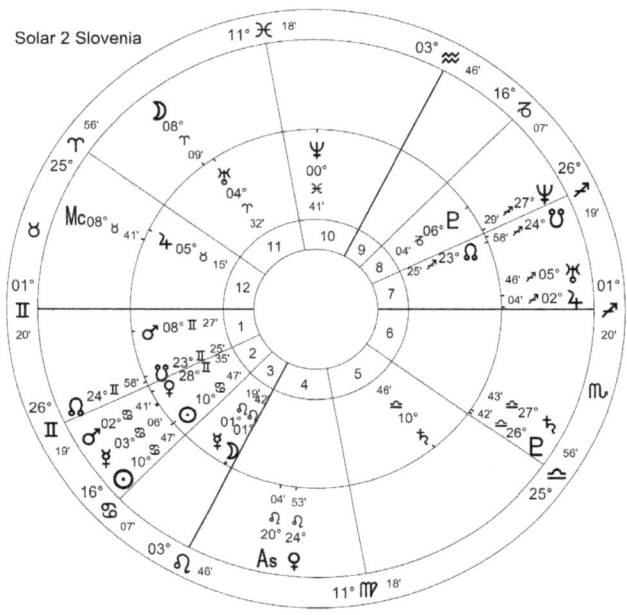

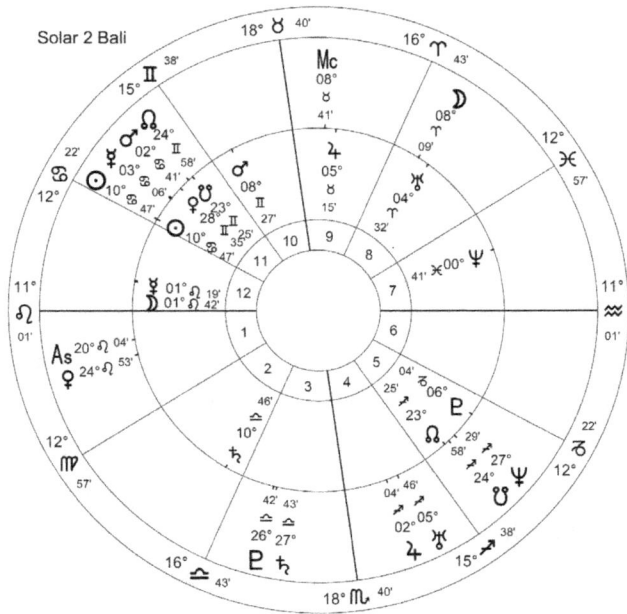

The two horoscopes, calculated for both places for the next year, show the difference between the two potentials, the person in question cannot be aware of, of course, having lived only one of the two options. At home, namely, this person would have been subject to conflicts and financial tensions, whereas the activity axis in the relocated chart focuses on the 11th and 5th houses, highlighting creativity and future. In addition, the Ascendant in the Ljubljana chart is in Gemini, with Mercury not very favourably placed, "dreaming of changes to come", while in the relocated chart the Ascendant is in Leo, with the Sun as part of a vision already being put into action at the time. In fact, this lady quickly found a sphere of activity much more to her liking than her previous occupation, namely spirituality and personal growth, conditions for these having been much better in Bali, where she continues to live.

It should, however, be pointed out here that this person did not depart for Bali after having consulted an astrologer about favourable potential destinations, but had simply followed her intuition. Astrology only entered her life later...

„Happy Birthday" on a Hawaiian beach

If we now know that a horoscope, including predictions, "behaves" differently in case of relocation, with a solar return horoscope only being valid for one year, we are not far from the idea that the relocation principle could be taken into account in our calculations. To put it simply: if we spend our birthday in a different location and calculate our solar return for that very location, we get a horoscope, different from the one, valid for our place of residence. The new, relocated annual horoscope brings information about different potentials, and with potentials differing with regard to places, the principle should theoretically work also in cases of short-term relocations.

Moreover, several astrologers argue that relocating at the time, crucial for calculating the solar return, affects the quality of our life in the respective year to such an extent, that finding a favourable location for a solar return makes sense.

This in turn generates a whole new art of finding locations around the world, where planets are favourably placed for career, partnership and - most often - health. Of course not all planets can be placed in favourable positions in this way, and a thorough consideration is needed to decide, whether a client will benefit more from, say, a well-placed Mars, than from an unfavourable Uranus, which cannot be avoided. The decision depends primarily on natal potentials, as well as on the desires and expectations of the client. However, proceeding from the well-known principles of electional astrology, referring to selection of the most appropriate date for an action or event, we know that it is possible to find an astrological chart to resonate with client's expectations. The essential difference in the relocation of a yearly horoscope is simply that we are not searching for the right time, but for the right location.

Is unambiguous answer even possible?

The practice of "choosing the right place for a birthday celebration" is rather widespread, and many people do indeed travel the world to improve their potential in the year ahead.

There are a number of relevant indicators, including angularity of planets and focus on angular versus cadent houses, depending on requirements and expectations.

Anthony Louis, an expert on solar returns, cites a number of approaches by various astrologers throughout history, beginning with the Frenchmen Morinus (Jean-Baptiste Morin de Villefranche, 1583-1656) and Alexandre Volguine (1903-1976). Louis also notes that "unlike Morinus, Penfield (Marc Penfield, 1942-) always calculates solar returns for the place of residence, irrespective of the person's actual location at the time of the return". Moreover, Penfield "firmly believes that a solar return chart should always be calculated for the place, where one lives, has a fixed address, pays rent and receives mail".

Marion March and Joan McEvers, authors of a series of textbooks on astrology, argue that "a solar return horoscope shows the pattern of daily life in the city of residence, not the place of birth". However, with regard to a birthday relocation - on the basis of the numerous studies they have carried out - "it cannot be claimed that the principle really works". I quote: *"Regardless of a holiday or business trip at the time of your solar return, our experience suggests that the solar return horoscope will continue to refer to the place you call 'home'."*

Astrologers mostly disagree on the issue of relocation, the latter being just another of the "great" dilemmas in astrology. Let us have a look at an illustrative example.

Moving across the Atlantic

In the early years of this century, Miss Marjana moved to Newark in the New York area. Shortly before that, her son had moved to Colorado to study, and after having completed his education, he moved in with her. As she had been a regular client of mine for a decade before, I had a whole series of solar return horoscopes for both, herself and her son in my archives. Between 2004 and 2013 we kept in touch at a distance and accumulated a lot of feedback information. In line with the relocation dilemma, ever since their move I have been making the calculations in parallel for Newark and Ljubljana, where they had lived before. The difference between the two locations is six time zones, meaning that each pair of horoscopes for a given year differed so in terms of axis orientation as in terms of houses. This meant, however, that for the most part it was easy to see, which of the two parallel horoscopes was valid, the Ljubljana chart or the relocated one.

Another question was raised. If we move to a distant place, we do so within a solar year, sometime between two birthdays. But even a relocation on, say, a birthday, should be visible in one of the two solar horoscopes. Is this supposed to mean that the initial solar horoscope, in this case the Ljubljana chart, is only valid until the day of departure, while the relocated horoscope applies to the rest of the solar year?

According to the theory, one would therefore expect that for both, the mother and the son, the two solar returns for 2004/2005, in which the relocation date, 15 January 2005, falls, would be calculated for the previous locations, i.e. for him Colorado and for her Ljubljana, while for Newark they would only be calculated from their birthdays in 2005, i.e. in May and March of that year.

However, the use of relocation did not work for the two solar returns of the son, because information about both years of living and schooling in Colorado could only be drawn from the solar return horoscopes, calculated for Ljubljana. Given his young age and the temporary validity of the location, this seemed quite logical. But soon, already in the first years of living in the American East, as well as in the first astrological consultations from this location, it became clear that the horoscopes were not as accurate as expected. And what is more: in the mother's case, the use of the relocated horoscope proved to be useful for all these years, while in the son's case, only the initial one, calculated for Ljubljana, seemed to be working!

On the other hand, the relocation is clearly visible in the mother's annual horoscope for Ljubljana, while in the next year, two important events, namely arrangements with relatives, financial problems and the associated fear are very clearly reflected in the **relocated** horoscope, as are several events from the years that followed. In the son's horoscope, however, the relocated solar returns did not even show important events like the great success in the sporting field, the move to Newark, problems in the new team, which later affected his sporting career, the graduation from college and the break-up of his relationship with his girlfriend, and, last but not least, not even the two changes of residence in a two-month period in the spring of 2009. Moreover, the latter events are quite clearly indicated in his annual horoscope for Ljubljana, although he had not lived there for more than five years!

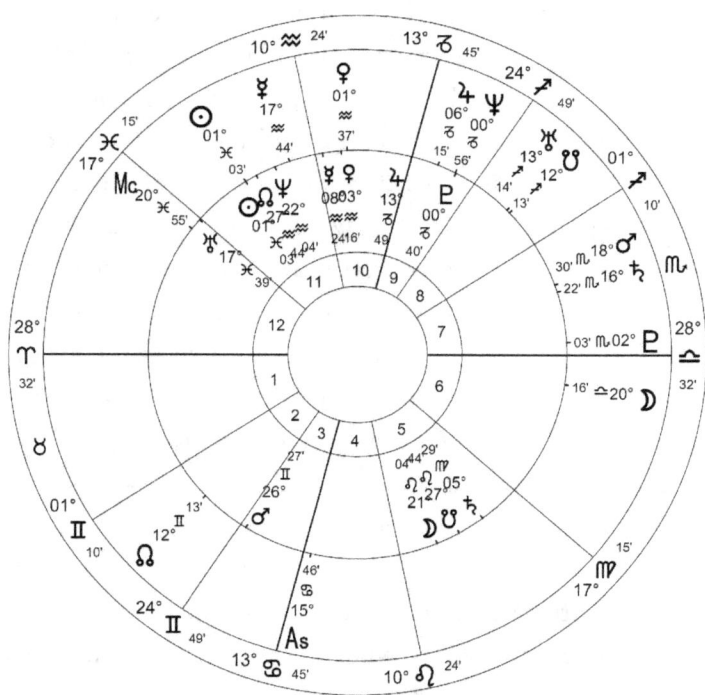

Jupiter, ruler of the 9th house, denoting studies, conjuncts the MC, bringing "a year when study and success come together", i.e. a year of life-changing success. But in terms of timing the planets confirm that he had made all the arrangements for his graduation around his birthday, and then waited until May to graduate, by which time other accumulated pressures suppressed the excitement of graduation, so the period is more notable for Pluto in the 9th house, natal Mercury in the 11th, and the Moon in the 5th. Even more noticeable, however, are the two potentials in the 5th and 7th houses, where differences of character brought about dissolution of the partnership in October, with the natal planets in the 7th house being pushed back a little in time – having had neither support nor affirmation, but rather more and more conflicts, he broke up the partnership himself (Saturn in the 5th house).

The logical conclusion is that even though mother and son lived in the same house for several years, he was impacted by a solar return horoscope for Ljubljana, and she by a solar return for Newark!

Am I now suddenly American (Hawaiian, French)?

This was not the only practical example in my yearlong astrology practice that made me sceptical about searching for a suitable birthday location for the sake of better potentials in the year to come; however, it is certainly the most thoroughly researched and also the most illustrative one. Neither the first nor the second option can be chosen on the basis of this case. What then is valid?

It is true that one feels at home in one place and not so in another. The question is whether we can feel "at home" at a location of a short-term stay, for the purpose of, say, participation at an international congress. Probably not. One returns to one's home place only to leave again, remaining forever a resident of the place of one's roots. Does it make a difference if a short stay in a foreign place is experienced just for our birthday?

From here on, there is probably no rule. To help us understand, perhaps we could refer to a psychological phenomenon from medicine, known as 'placebo'. It is related to the **belief** that a relocation can change our potentials, but there is no question of the principle's accuracy. It certainly works, but the question is for whom, or in which case.

I remember a case where a client returned to her native country on short notice just a month before her birthday, because her mother had died. This emotional shock was clearly visible in the relocated horoscope made for Basel, where she had lived and worked for many years. All the important activities, however, relating not only to the consequences of this loss, but also to finding a new job and to administrative procedures, were clearly reflected in the corresponding solar return, calculated for her birthplace in Slovenia, where she had returned!

And another example: for a client with a very changeable and flexible character and horoscope, who uses her potential perfectly in her profession and career, I have been doing calculations for Maribor, where she was born, for many years, even though she has been living and working in Norway for the last eight years, and had even lived in Ljubljana for several consecutive years before...

One way or another

In my work, I have been trying to trace a pattern or common characteristics, astrological indicators, showing the rules of application of a basic solar return versus a relocated one. But even after many examples, the dilemma remains unresolved. In the aforementioned case there are, in my humble opinion, two factors that may be decisive. The first one is age and perhaps also the flexibility that goes with it, however unpleasant this may sound. But then one could expect the younger generation to be faster in 'switching' its perception, so the principle of relocation should have been applied to the son. But it was the other way around.

Another factor is the astrological chart or character assessment. Some people are more flexible than others, or on the other hand, more stable, constant, perhaps more stubborn. In our case the mother, a Gemini, has the Sun/Moon conjunction in the tenth house, while the son has the Sun in Pisces and in the ninth house (foreign countries), but also Cancer Ascendant with Moon in Libra and in the fourth house, related to the sign of Cancer homeland and tradition. Could this be the prevailing factor in the assessment of applicability of a relocated horoscope? Possibly, but according to practical examples that mostly depends upon individual horoscopes.

In several cases it was necessary to use relocated solar returns for people having lived abroad for a long time. Is it a matter of slowly beginning to "belong here" or "this being my home now"? It would be interesting to determine the factors and circumstances bringing about such inner change...

A few more tidbits

If we wanted to get to the bottom of the relocation dilemma, we could consider some more extensive research at this point, including various borderline cases, such as when someone spends the day for which the solar return horoscope is calculated - not necessarily their birthday -, somewhere else, and then returns home "in time" to celebrate with family and relatives. What happens if a person is not aware of having been elsewhere "on that very day"? Or, say, the case where someone decides to move to a calculated and presumably more appropriate location, but arrives too late due to transportation problems, or spends a calculated birthday moment on a transatlantic flight? My own professional scepticism on this issue is, to be honest, largely due to the fact that I work in a small country, whose population is characterized by infrequent, let alone distant relocating. To be honest, we cannot even move very far within this country - two hundred kilometres from its centre is already a cross-border territory on all sides... Obviously I have been less focused on this astrological dilemma than someone living, for example, in the United States, where people often move from one end to another.

A few small but useful observations are worth mentioning here. People move in all sorts of directions, and astrologers will be particularly interested in the east-west and north-south variations, since latitude and longitude are the most important calculation factors. Here it can be observed that latitudinal moves do not produce as marked a change in the calculation as moving east or west.

In my own research, I have made a whole series of horoscopes for a particular moment in time, based on the 15th degree of eastern latitude, this being the latitude going through Slovenia.

This meridian embraces the Spitzberg archipelago, Stargard in Poland, Görlitz in Germany, Trebnje in Slovenia, Mount Etna and Ndjamena in Chad.

Had we therefore moved just a few hundred kilometres north or south of our starting point, i.e. our place of residence, the relocated horoscope would not be very different from the initial one. Let us compare two places in Europe at exactly the same distance, 619 kilometres, from the starting point of the calculation, Ljubljana, namely Dresden in Germany and Monaco. Their latitudes differ from that of Ljubljana by 5 and just over 2 degrees respectively, and their longitudes by 7 and 1 degrees respectively. The solar return Ascendants for the same year are Ljubljana - 16°16 Leo, Dresden - 17°59 Leo and Monaco - 9°38 Leo. (Anyone can play with these calculations, provided they have the right astrology software and a slightly better map.)

Bottom line: the idea that selection of other places has a strong and significant impact on life, highlighting qualities that may be less pronounced, little noticed or suppressed in the place of origin, is logical and useful; however, I am rather sceptical about the value of visiting distant places for a short time for one's birthday with regard to the quality of life in the year that follows.

XII
CONCLUSION

A Solar Return Horoscope is a unique astrological forecasting technique offering an overview of activity in the year between two birthdays. Having read the book you probably realized that the presented approach to the solar return method is largely based on practice and insights accumulated over thirty years of work in the astrological field.

This technique offers a wealth of useful information on activity and potentials in the year ahead, as well as on the processes and development of one's personality, especially in combination with birth horoscope as a map of individual's life. Based on my personal approach to timing within a solar year, the solar return method has proven to be a simple and very useful predictive technique. Its results should, according to the currently valid rules of astrological prediction, be combined with other methods if we are to be sure about the credibility of a solar return chart, however, it is perfectly possible to use solar return as the main predictive method, not merely as an additional one, as claimed by several authors. What is more, the method provides for the kind of narrative that comes closer to the comprehension ability of astrologer's clients. Using a psychological approach, individual parts of yearly horoscope can be linked not only into a traceable series of events, but also into a logically coherent whole, offering client an insight into the development of his situation. Generating a series of solar returns for several years in advance offers an opportunity to determine the best time to initiate an activity or complete a project, while reasons for different delays should be sought in specific interconnections of individual factors of a horoscope.

Unlike other techniques, which focus on actual or symbolic placements of the planets at a given time, making one feel like a pawn on the chessboard of planets, powers and gods, a solar return largely reflects human perception, not just events as they actually occur. Due to this fact, the technique shows its greatest value in cases where we, as consultants, work with an individual who participates with, reports and assists astrologer with feedback and other information. In this way it is possible to evaluate in a useful and practical way the events that have taken place and are yet to unfold.

This makes it easier to answer the question of adequacy of solar returns. The question is whether findings and indications in solar returns can be applied to young children; personally I do not use this method with children under the age of 15, and from then on until coming of age, according to ethical principles, only in the presence of or with a written consent of parents.

There is also the question of applicability of solar returns to different entities, such as companies or countries. Above all, there is the dilemma of who can identify with this horoscope, or who is the subject perceiving events, foreseen in the solar return horoscope.

A president, director, government, population, employees? I use my technique with utmost care in such cases, quite differently from working with individuals, where it is in many ways unsurpassable.

The presented technique of solar returns has also proven to be very useful in cases of relocation, rectification and to some extent even reconstruction of events or developments. The latter applies both to past events as to future trends.

Perhaps the greatest advantage of the solar return method, presented in this book, is in the fact that neither transiting nor progressed planets need to be inserted in the horoscope wheel to determine, **when in the year events indicated by solar return chart are to unfold**, all information being provided by the return chart itself. However, a combined chart with natal horoscope offers the possibility of examining several areas in the same year and even simultaneously within the same period, thus allowing for integration of different life spheres that would otherwise have to be dealt with separately, using other techniques.

A solar return horoscope provides us with ample information in a simple, direct and efficient way.

Bibliography:

Aubier, Catherine: *"Encyclopedia of Astrology"* (DZS, Ljubljana, 1991)

Chevalier, Jean, Alain Gheerbrant: *"Dictionary of Symbols"* (Mladinska knjiga, Ljubljana, 1995)

Discepolo, Ciro: *"Transiti e rivoluzioni solari"* (Gruppo Editoriale Armenia, 1997)

Eshelman, James A.: *"Interpreting Solar Returns"* (1979; Serbian edition: Dečje novine, Gornji Milanovac, 1991)

Geddes, Sheila: *"The Art of Astrology"* (The Aquarian Press, 1980)

Kirby, Babs, Janey Stubbs: *"Interpreting Solar & Lunar Returns: A Psychological Approach"*; Serbian edition: Metaphysica, Belgrade, 2006.

Louis, Anthony: *"The Art of Forecasting using Solar Returns"* (The Wessex Astrologer, 2008)

March, Marion, Joan McEvers: *"The only way to ... Learn Tomorrow, Vol. IV, Current Patterns"*; Serbian edition.)

Merriman, Raymond E.: *"The "New" Solar Return Book of Prediction"* (Seek-It Publications, 9th Printing, 1998,)

Rose, Deanna: *"Easy Predictions with Solar Returns"* (21st Century Metaphysical Publishing, fourth revised edition, 1998)

"Dictionary of the Slovene Literary Language" (DZS, handy reprint in 15 books, 2008)

Šalamun, Ivana: *"Astrology, the game of planets and stars"* (in Slovenian; Borec Publishing House, Ljubljana, 1990)

Teal, Celeste: *"Identifying Planetary Triggers"* (Llewellyn Publications, St. Paul, 2000)

Teal, Celeste: *"Predicting Events with Astrology"* (Llewellyn Publications, Woodbury, 2009)

Teal, Celeste: *"Lunar Nodes*: Discover Your Soul's Carmic Mission" (published in Slovenian by Anu elara, Ljubljana, 2013)

Appendix 1

SOME EMPHASIS AND GROUPING IN SOLAR RETURN CHARTS

Axes	
Focus on the left side	Year of decisions
Focus on the right side	Less control over activity
Focus on the upper half	External activity, ambition, objectivity
Focus on the lower half	Passivity, introversion, inner motives, subjectivity
Chart segments (house groupings)	
1 – 4	Focus on self
5 – 8	Focus on others
9 – 12	Business, socializing, charity, growth
Angular houses (1., 4., 7. in 10.)	New projects and/or relationships, beginnings
Succedent houses (2., 5., 8. in 11.)	Avoiding or refusing changes, stability
Cadent houses (3., 6., 9. in 12.)	Year of changes, spiritual activity

SOME TYPICAL HOUSE COMBINATIONS IN SOLAR RETURN CHARTS

Topic	Activated astrological houses
Marriage	03, 07, 09, 11
Child	04, 05, 03 (also 09)
Change of job	02, 06, 10
Relocation	03, 04, 06
Divorce	07, 09, 03, 08
Health	06, 08, 12, (01)
Inheritance	08, 02
Gambling	05, 08, 02
Love:	
- love affair	05, 07, 11, VE
- affair with a married man	05, 12
- partner's affair	07, 12, MA

Appendix 2

Dividing the astrological houses into twelfths

With the help of these tables, the reader will more easily help himself in determining the time of the planetary position within a certain astrological house of the solar return horoscope.

The range of the astrological house from 12 to 72 degrees is taken into account, which can mostly be found in calculated solar charts for a rather wide geographical area (latitude).

In the first table we find only the calculation of the range of one month of the year, while the second table gives the same calculation for all twelve months of the year for which the solar return is calculated.

deg.	1/12 =
12	1
13	1 1/12
14	1 1/6
15	1 1/4
16	1 1/3
17	1 5/12
18	1 1/2
19	1 7/12
20	1 2/3
21	1 3/4
22	1 5/6
23	1 11/12
24	2
25	2 1/12
26	2 1/6
27	2 1/4
28	2 1/3
29	2 5/12
30	2 1/2
31	2 7/12
32	2 2/3

deg.	1/12 =
33	2 3/4
34	2 5/6
35	2 11/12
36	3
37	3 1/12
38	3 1/6
39	3 1/4
40	3 1/3
41	3 5/12
42	3 1/2
43	3 7/12
44	3 2/3
45	3 3/4
46	3 5/6
47	3 11/12
48	4
49	4 1/12
50	4 1/6
51	4 1/4
52	4 1/3

deg.	1/12 =
53	4 5/12
54	4 1/2
55	4 7/12
56	4 2/3
57	4 3/4
58	4 5/6
59	4 11/12
60	5
61	5 1/12
62	5 1/6
63	5 1/4
64	5 1/3
65	5 5/12
66	5 1/2
67	5 7/12
68	5 2/3
69	5 3/4
70	5 5/6
71	5 11/12
72	6

Appendix 2

deg.	1 M	2 M	¼ Y	4 M	5 M	½ Y	M	8. M	¾ Y	10 M	11 M	1 Y
12	1°	2°	3°	4°	5°	6°	7°	8°	9°	10°	11°	12°
13	1 1/12	2 1/6	3 1/4	4 1/3	5 5/12	6 1/2	7 7/12	8 2/3	9 3/4	10 5/6	11 11/12	13
14	1 1/6	2 1/3	3 1/2	4 2/3	5 5/6	7	8 1/6	9 1/3	10 1/2	11 2/3	12 5/6	14
15	1 1/4	2 1/2	3 3/4	5	6 1/4	7 1/2	8 3/4	10	11 1/4	12 1/2	13 3/4	15
16	1 1/3	2 2/3	4	5 1/3	6 2/3	8	9 1/3	10 2/3	12	13 1/3	14 2/3	16
17	1 5/12	2 5/6	4 1/4	5 2/3	7 1/12	8 1/2	9 11/12	11 1/3	12 3/4	14 1/6	15 7/12	17
18	1 1/2	3	4 1/2	6	7 1/2	9	10 1/2	12	13 1/2	15	16 1/2	18
19	1 7/12	3 1/6	4 3/4	6 1/3	7 11/12	9 1/2	11 1/12	12 2/3	14 1/4	15 5/6	17 5/12	19
20	1 2/3	3 1/3	5	6 2/3	8 1/3	10	11 2/3	13 1/3	15	16 2/3	18 1/3	20
21	1 3/4	3 1/2	5 1/4	7	8 3/4	10 1/2	12 1/4	14	15 3/4	17 1/2	19 1/4	21
22	1 5/6	3 2/3	5 1/2	7 1/3	9 1/6	11	12 5/6	14 2/3	16 1/2	18 1/3	20 1/6	22
23	1 11/12	3 5/6	5 3/4	7 2/3	9 7/12	11 1/2	13 5/12	15 1/3	17 1/4	19 1/6	21 1/12	23
24	2	4	6	8	10	12	14	16	18	20	22	24
25	2 1/12	4 1/6	6 1/4	8 1/3	10 5/12	12 1/2	14 7/12	16 2/3	18 3/4	20 5/6	22 11/12	25
26	2 1/6	4 1/3	6 1/2	8 2/3	10 5/6	13	15 1/6	17 1/3	19 1/2	21 2/3	23 5/6	26
27	2 1/4	4 1/2	6 3/4	9	11 1/4	13 1/2	15 3/4	18	20 1/4	22 1/2	24 3/4	27
28	2 1/3	4 2/3	7	9 1/3	11 2/3	14	16 1/3	18 2/3	21	23 1/3	25 2/3	28
29	2 5/12	4 5/6	7 1/4	9 2/3	12 1/12	14 1/2	16 11/12	19 1/3	21 3/4	24 1/6	26 7/12	29
30	2 1/2	5	7 1/2	10	12 1/2	15	17 1/2	20	22 1/2	25	27 1/2	30
31	2 7/12	5 1/6	7 3/4	10 1/3	12 11/12	15 1/2	18 1/12	20 2/3	23 1/4	25 5/6	28 5/12	31
32	2 2/3	5 1/3	8	10 2/3	13 1/3	16	18 2/3	21 1/3	24	26 2/3	29 1/3	32
33	2 3/4	5 1/2	8 1/4	11	13 3/4	16 1/2	19 1/4	22	24 3/4	27 1/2	30 1/4	33
34	2 5/6	5 2/3	8 1/2	11 1/3	14 1/6	17	19 5/6	22 2/3	25 1/2	28 1/3	31 1/6	34
35	2 11/12	5 5/6	8 3/4	11 2/3	14 7/12	17 1/2	20 5/12	23 1/3	26 1/4	29 1/6	32 1/12	35
36	3	6	9	12	15	18	21	24	27	30	33	36
37	3 1/12	6 1/6	9 1/4	12 1/3	15 5/12	18 1/2	21 7/12	24 2/3	27 3/4	30 5/6	33 11/12	37
38	3 1/6	6 1/3	9 1/2	12 2/3	15 5/6	19	22 1/6	25 1/3	28 1/2	31 2/3	34 5/6	38
39	3 1/4	6 1/2	9 3/4	13	16 1/4	19 1/2	22 3/4	26	29 1/4	32 1/2	35 3/4	39
40	3 1/3	6 2/3	10	13 1/3	16 2/3	20	23 1/3	26 2/3	30	33 1/3	36 2/3	40
41	3 5/12	6 5/6	10 1/4	13 2/3	17 1/12	20 1/2	23 11/12	27 1/3	30 3/4	34 1/6	37 7/12	41
42	3 1/2	7	10 1/2	14	17 1/2	21	24 1/2	28	31 1/2	35	38 1/2	42
43	3 7/12	7 1/6	10 3/4	14 1/3	17 11/12	21 1/2	25 1/12	28 2/3	32 1/4	35 5/6	39 5/12	43
44	3 2/3	7 1/3	11	14 2/3	18 1/3	22	25 2/3	29 1/3	33	36 2/3	40 1/3	44
45	3 3/4	7 1/2	11 1/4	15	18 3/4	22 1/2	26 1/4	30	33 3/4	37 1/2	41 1/4	45
46	3 5/6	7 2/3	11 1/2	15 1/3	19 1/6	23	26 5/6	30 2/3	34 1/2	38 1/3	42 1/6	46
47	3 11/12	7 5/6	11 3/4	15 2/3	19 7/12	23 1/2	27 5/12	31 1/3	35 1/4	39 1/6	43 1/12	47
48	4	8	12	16	20	24	28	32	36	40	44	48
49	4 1/12	8 1/6	12 1/4	16 1/3	20 5/12	24 1/2	28 7/12	32 2/3	36 3/4	40 5/6	44 11/12	49
50	4 1/6	8 1/3	12 1/2	16 2/3	20 5/6	25	29 1/6	33 1/3	37 1/2	41 2/3	45 5/6	50
51	4 1/4	8 1/2	12 3/4	17	21 1/4	25 1/2	29 3/4	34	38 1/4	42 1/2	46 3/4	51
52	4 1/3	8 2/3	13	17 1/3	21 2/3	26	30 1/3	34 2/3	39	43 1/3	47 2/3	52
53	4 5/12	8 5/6	13 1/4	17 2/3	22 1/12	26 1/2	30 11/12	35 1/3	39 3/4	44 1/6	48 7/12	53

SOLAR RETURNS

54	4 1/2	9	13 1/2	18	22 1/2	27	31 1/2	36	40 1/2	45	49 1/2	54
55	4 7/12	9 1/6	13 3/4	18 1/3	22 11/12	27 1/2	32 1/12	36 2/3	41 1/4	45 5/6	50 5/12	55
56	4 2/3	9 1/3	14	18 2/3	23 1/3	28	32 2/3	37 1/3	42	46 2/3	51 1/3	56
57	4 3/4	9 1/2	14 1/4	19	23 3/4	28 1/2	33 1/4	38	42 3/4	47 1/2	52 1/4	57
58	4 5/6	9 2/3	14 1/2	19 1/3	24 1/6	29	33 5/6	38 2/3	43 1/2	48 1/3	53 1/6	58
59	4 11/12	9 5/6	14 3/4	19 2/3	24 7/12	29 1/2	34 5/12	39 1/3	44 1/4	49 1/6	54 1/12	59
60	5	10	15	20	25	30	35	40	45	50	55	60
61	5 1/12	10 1/6	15 1/4	20 1/3	25 5/12	30 1/2	35 7/12	40 2/3	45 3/4	50 5/6	55 11/12	61
62	5 1/6	10 1/3	15 1/2	20 2/3	25 5/6	31	36 1/6	41 1/3	46 1/2	51 2/3	56 5/6	62
63	5 1/4	10 1/2	15 3/4	21	26 1/4	31 1/2	36 3/4	42	47 1/4	52 1/2	57 3/4	63
64	5 1/3	10 2/3	16	21 1/3	26 2/3	32	37 1/3	42 2/3	48	53 1/3	58 2/3	64
65	5 5/12	10 5/6	16 1/4	21 2/3	27 1/12	32 1/2	37 11/12	43 1/3	48 3/4	54 1/6	59 7/12	65
66	5 1/2	11	16 1/2	22	27 1/2	33	38 1/2	44	49 1/2	55	60 1/2	66
67	5 7/12	11 1/6	16 3/4	22 1/3	27 11/12	33 1/2	39 1/12	44 2/3	50 1/4	55 5/6	61 5/12	67
68	5 2/3	11 1/3	17	22 2/3	28 1/3	34	39 2/3	45 1/3	51	56 2/3	62 1/3	68
69	5 3/4	11 1/2	17 1/4	23	28 3/4	34 1/2	40 1/4	46	51 3/4	57 1/2	63 1/4	69
70	5 5/6	11 2/3	17 1/2	23 1/3	29 1/6	35	40 5/6	46 2/3	52 1/2	58 1/3	64 1/6	70
71	5 11/12	11 5/6	17 3/4	23 2/3	29 7/12	35 1/2	41 5/12	47 1/3	53 1/4	59 1/6	65 1/12	71
72	6	12	18	24	30	36	42	48	54	60	66	72

www.ingramcontent.com/pod-product-compliance
Lightning Source LLC
Chambersburg PA
CBHW071018240526
45469CB00006BD/1973